AF412842

102
Structure and Bonding

Managing Editor:
D.M.P. Mingos

Springer
Berlin
Heidelberg
New York
Barcelona
Hong Kong
London
Milan
Paris
Tokyo

High Performance
Non-Oxide Ceramics II

Volume Editor: M. Jansen

With contributions by
R. Haubner, M. Herrmann, B. Lux, G. Petzow,
R. Weissenbacher, M. Wilhelm

Springer

The series *Structure and Bonding* publishes critical reviews on topics of research concerned with chemical structure and bonding. The scope of the series spans the entire Periodic Table. It focuses attention on new and developing areas of modern structural and theoretical chemistry such as nanostructures, molecular electronics, designed molecular solids, surfaces, metal clusters and supramolecular structures. Physical and spectroscopic techniques used to determine, examine and model structures fall within the purview of Structure and Bonding to the extent that the focus is on the scientific results obtained and not on specialist information concerning the techniques themselves. Issues associated with the development of bonding models and generalizations that illuminate the reactivity pathways and rates of chemical processes are also relevant.

As a rule, contributions are specially commissioned. The editors and publishers will, however, always be pleased to receive suggestions and supplementary information. Papers are accepted for *Structure and Bonding* in English.

In references *Structure and Bonding* is abbreviated *Struct Bond* and is cited as a journal.

Springer WWW home page: http://www.springer.de
Visit the SB home page at http://link.springer.de/series/sb/ or
http://link.springer-ny.com/series/sb/

ISSN 0081-5993
ISBN 3-540-43132-2
Springer-Verlag Berlin Heidelberg New York

CIP Data applied for

Springer-Verlag Berlin Heidelberg New York a member of BertelsmannSpringer
Science + Business Media GmbH
http://www.springer.de
© Springer-Verlag Berlin Heidelberg 2002
Printed in Germany

Typesetting: Scientific Publishing Services (P) Ltd, Madras
Production editor: Christiane Messerschmidt, Rheinau
Cover: Medio Technologies AG, Berlin
Printed on acid-free paper   SPIN: 108 57239   02/3020 – 5 4 3 2 1 0

Structure and Bonding
Also Available Electronically

For all customers with a standing order for Structure and Bonding we offer the electronic form via LINK free of charge. Please contact your librarian who can receive a password for free access to the full articles by registration at:

http://link.springer.de/orders/index.htm

If you do not have a standing order you can nevertheless browse through the table of contents of the volumes and the abstracts of each article at:

http://link.springer.de/series/sb/
http://link.springer-ny.com/series/sb/

There you will also find information about the

- Editorial Board
- Aims and Scope
- Instructions for Authors

Preface

The nitrides and carbides of boron and silicon are proving to be an excellent choice when selecting materials for the design of devices that are to be employed under particularly demanding environmental and thermal conditions. The high degree of cross-linking, due to the preferred coordination numbers of the predominantly covalently bonded constituents equalling or exceeding three, lends these non-oxidic ceramics a high kinetic stability, and is regarded as the microscopic origin of their impressive thermal and mechanical durability. Thus it does not come as a surprise that the chemistry, the physical properties and the engineering of the corresponding binary, ternary, and even quaternary compounds have been the subject of intensive and sustained efforts in research and development.

In the five reviews presented in the volumes 101 and 102 of "Structure and Bonding" an attempt has been made to cover both the essential and the most recent advances achieved in this particular field of materials research. The scope of the individual contributions is such as to address both graduate students, specializing in ceramic materials, and all scientists in academia or industry dealing with materials research and development. Each review provides, in its introductory part, the chemical, physical and, to some extent, historical background of the respective material, and then focuses on the most relevant and the most recent achievements.

Since the degree of maturity reached by the materials considered is rather varied, the focus of the respective reviews is also quite different. Thus for SiC and Si_3N_4, the main emphasis is placed on processing and shaping, while for BN its transformation to the cubic polymorph is a major concern, and, finally, the report on the still rather young class of amorphous Si/B/N/C ceramics is mainly devoted to aspects related to chemical syntheses and basic characterizations. Inspite of the fact that many phenomena are dominated by kinetic control, knowing the underlying thermodynamic equilibria is a crucial prerequisite to any deeper understanding of the nitride and carbide based materials discussed here. Therefore, a comprehensive and critically assessed compilation of thermodynamic data and phase equilibria for the quaternary system Si/B/N/C as well as its ternary and binary sub-systems has been included as an introductory chapter preceding the reviews devoted to specific materials and their properties.

Stuttgart, April 2002   Martin Jansen

Contents

Boron Nitrides – Properties, Synthesis and Applications

R. Haubner[1], M. Wilhelm[2], R. Weissenbacher[2], B. Lux[2]

[1] *e-mail: rhaubner@mail.zserv.tuwien.ac.at*
Tel.: +43 1 58801 16128, Fax: +43 1 58801 16199
[2] Institute for Chemical Technology of Inorganic Materials, University of Technology Vienna, Getreidemarkt 9/161, 1060 Vienna, Austria

Boron nitride is a extraordinary topic in the area of materials science. Due to the special bonding behaviors of boron and nitrogen the BN exists in many different structures. The well-defined crystallographic structures are hexagonal BN (h-BN), rhombohedral BN (r-BN), wurtzitic BN (w-BN), and cubic BN (c-BN). Additionally, other crystalline and amorphous structures exist. Exceptional is that there are still discussions about the BN phase diagram. In the present stage c-BN is the stable phase at standard conditions but exact data about the phase transition line are not yet available. Synthesis of h-BN powders and coatings is described as well as applications of BN in ceramic materials and as lubricant. For c-BN the high-pressure high-temperature synthesis for powder production is discussed, and an overview about applications in wear resistant ceramics (polycrystalline c-BN) is given. The low-pressure methods for nano-cBN deposition (PVD and Plasma CVD) are described.

Keywords: Boron nitride, Hexagonal-BN, Cubic-BN, High-pressure high-temperature synthesis, Chemical vapor deposition

Structure and Bonding, Vol. 102
© Springer-Verlag Berlin Heidelberg 2002

List of Abbreviations and Symbols

a-BN	amorphous BN
BN	boron nitride
c-BN	cubic BN
CVD	chemical vapor deposition
ΔG	free enthalpy
ΔH	enthalpy
ΔS	entropy
d.c.	direct current
DTA	differential thermo analysis
E-BN	BN synthesized by explosion
ECR	electron cyclotron resonance
FTIR	Fourier transformation infrared spectroscopy
g-BN	graphitic BN
h-BN	hexagonal BN
HIP	hot-isostatic press
HP	hot-pressing
HP-HT	high-pressure high-temperature
IBAD	ion-beam-assisted deposition
IBD	ion beam deposition
i-BN	BN prepared by energetic ions
ICP	inductively coupled plasma
IR	infrared spectroscopy
LED	light emitting diode
LO	longitudinal optical mode

MISFET   metal insulator semiconductor field effect transistor
nano-cBN   nano-crystalline c-BN
PACVD   plasma assisted CVD
PcBN   polycrystalline c-BN
PVD   physical vapor deposition
r-BN   rhombohedral BN
Ref.   reference
RF   radio-frequency
SEM   scanning electron microscope
t-BN   turbostratic BN
TO   transverse optical mode
w-BN   wurtzitic BN
XRD   X-ray diffraction

1
Introduction

Boron nitride (BN) – in all its various structures – is a synthetic product and not found in nature. The first synthesis for h-BN was described in 1842 by Balmain [1], but h-BN became a commercial material about 100 years later. Boron and nitrogen are neighbors of carbon in the periodic table, and therefore BN phases are isoelectric to the corresponding carbon phases. The structure similarities between h-BN and graphite have been well known, and therefore in 1957 Wentorf [2] successfully tried the high-temperature high-pressure synthesis of c-BN analogous to the diamond synthesis. The second hardest material – after diamond – was born, and since 1969 c-BN has been commercially available. Today c-BN is synthesized by various companies like General Electric (USA), De Beers (South Africa), Sumitomo and Showa Denko (Japan), and various companies in Russia.

Parts prepared of h-BN as well as c-BN are of great interest for industrial applications but also for materials science. The thermodynamic data for c-BN and the BN-phase diagrams found in literature are not in agreement. After the first high pressure experiments the B-N phase diagram was designed, and after some modifications c-BN was described as metastable phase at room temperature. Contrary to this opinion in 1988 it was reported that c-BN is the stable phase. Many experiments have confirmed this result, but exact thermodynamic data are still not available.

During the last few years the low pressure c-BN synthesis has been the most important topic in this field. Using PVD methods it has been possible to nucleate c-BN on various substrates, but only growth of nm-sized c-BN crystals is possible at the moment.

Due to its excellent properties h-BN is mainly used as ceramic material, as lubricant and serves also as thin coatings for electronic devices.

Utilization of cubic-BN are wear applications like machining tools and polishing powders.

2
Properties of the Various BN Phases

In this section we give a short review of various BN phases and their properties.

2.1
The BN Phases

Similar to the carbon system, BN exists in a soft hexagonal (h-BN) modification, a hard cubic (c-BN) one, and many others which are not very well crystallized, or amorphous. The properties of h-BN and c-BN are summarized in Table 1 [2–17], and the crystal structures of c-BN, w-BN (wurtzitic-BN), and h-BN are illustrated in Fig. 1.

Table 1. Properties of h-BN and c-BN

Property	Hexagonal BN	Cubic BN
Crystallographic data		
Crystal structure	Hexagonal	Cubic, zinc blende
Space group	$P6_3/mmc$ [3]	Fd3 m (8 atoms/unit cell)
Lattice constant	a = 2.504 Å [4]	a – 3.615 Å [2]; a – 3.67 Å [6]
	c = 6.661 Å [4]	a_{calc} = 3.606 Å [7]
B–N ion distance	1.446 Å [4]	1.57 Å [8]
Density	2.34 g/cm^3 [3]	3.4879 ± 0.003 g/cm^3 [9]
		3.45 g/cm^3 [2]
Mechanical properties		
Hardness	1.5–1.3 GPa (Vickers)	58–76 GPa (Knoop) [11]
	[10] (hot-pressed)	4500 kg/mm^2 [12]
Young's modulus	3400–8700 kg/mm^2 [13]	
Optical and electrical properties		
Color	White, gray	Colorless; B excess changes to yellow, orange, black [8]
Electrical resistivity	a/b axis 3.0×10^7 Ω cm [4]	10^{10} Ω · m (289 K) [8]
	c axis 3.0×10^9 Ω cm [4]	10^7 Ω · m (773 K) [8]
		10^{13} Ω · m [14]
		3.3×10^{13} Ω cm [15]
Bandgap	-	3.67 eV [6]
		6.4 ± 0.5 eV [16]
		5.0 eV [17]
Refraction index	-	2.117 (at 589.3 nm) [12]
		1.5–1.6 [17]
		Maximum 2.295 [15]
Thermal properties		
Thermal conductivity	a/b axis 0.627 W/cm · K [4]	13 W/cm · K [12]
	c axis 0.015 W/cm · K [4]	
Debye temperature		1700 K [5]
Linear thermal	a/b axis 3.24×10^{-1} K^{-1} [4]	4.80×10^{-1} K^{-1} (700 K) [8]
expansion	c axis 81×10^{-1} K^{-1} [4]	5.60×10^{-1} K^{-1} (1170 K) [8]
	(pyrolytic BN)	5.80×10^{-1} K^{-1} (1430 K) [8]

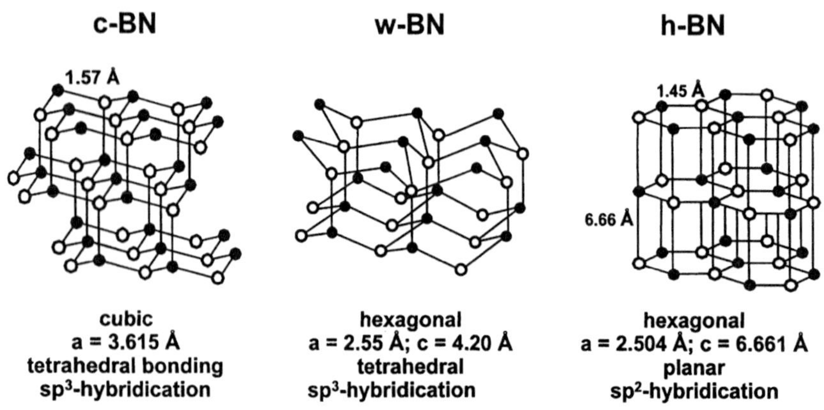

Fig. 1. Crystal structures of c-BN, w-BN, and h-BN

2.1.1
Hexagonal Boron Nitride (h-BN)

Hexagonal BN is mostly named h-BN, but also α-BN or g-BN (graphitic-BN) are used. It crystallizes similar to graphite in a hexagonal sheet layered structure, and therefore it is often referred to as "white graphite" (Fig. 1). The atomic planes are built by hexagonal rings formed by B and N atoms. The covalent bonds (σ-bonding, sp²-hybridization) between the atoms forming the rings are very strong. Between the atomic planes the bonding forces are weak, being van der Waals bonding (π-bonding). Additionally, it has to be pointed out that the planes are stacked on top of one another, without any horizontal displacement (boron and nitrogen are alternating along the c-axis). Due to the higher electronegativity of nitrogen the π-electron is located at the nitrogen and therefore h-BN is an electrical insulator and its color is white.

The morphologies of typical h-BN powders are shown in Fig. 2.

Fig. 2. Pictures of h-BN powders. (Courtesy of ESK-Kempten, a company of WACKER-Chemie)

2.1.2
Cubic Boron Nitride (c-BN) and Wurtzite-BN (w-BN)

Cubic boron nitride is commonly called c-BN in literature, but also z-BN (zinc blende) or β-BN [13] can be found. Wentorf [18] named c-BN "Borazon", which has become the trade name for the products of the General Electric Corporation. Russian companies call abrasive powders of c-BN "Elbor" or "Cubonite" [8].

Similar to the diamond lattice the B and N atoms are tetrahedrally coordinated. Every boron atom is surrounded by four nitrogen atoms and vice versa. In this arrangement boron and nitrogen atoms have sp^3 hybridization.

Because of the special bonding conditions (short bonding length) c-BN and diamond exhibit high hardness. Both materials are insulators because of missing π-bonds. The high thermal conductivity is caused by phonons and not by electrons like in metals.

Typical images of c-BN grown by the high-pressure high-temperature method are shown in Fig. 3.

Boron nitride can also form a superhard hexagonal phase in wurtzite-type (w-BN). This modification is a high pressure phase and was described first by Bundy and Wentorf [19].

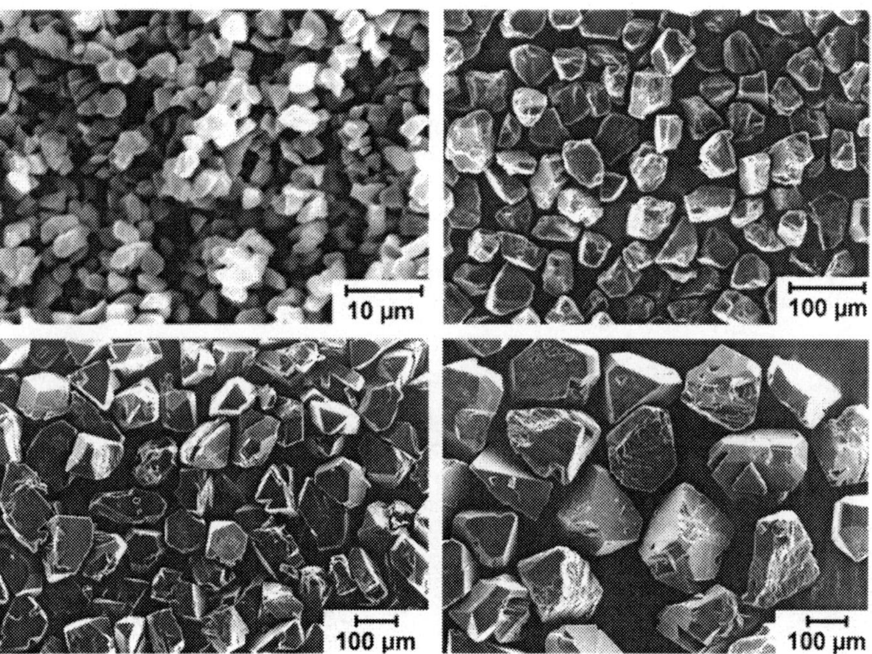

Fig. 3. Various commercially available c-BN powders with different grain size and morphology

Lattice parameters for w-BN are: a = 2.55 ± 0.01 Å; c = 4.20 ± 0.01 Å; and δ = 3.49 ± 0.03 g/cm^3 [19].

Similar data are found in later references [20, 21].

2.1.3
Rhombohedral and Turbostratic Structure

The rhombohedral (r-BN) structure is similar to the h-BN phase but the atomic layers sequence is ABC ABC. It was reported that r-BN is formed during conversion of c-BN into h-BN [22] (Fig. 4).

Turbostratic structure is characterized by a layer structure similar to h-BN; the layers are mostly parallel but not aligned to the c-axis [10] (Fig. 4).

2.1.4
Amorphous Structure

Amorphous boron nitride (a-BN) can be synthesized by decomposition of B-trichloroborazine and cesium [23]. Such layers are applied during production of semiconductor devices [24].

2.1.5
E-BN and i-BN

E-BN (E = explosion) is described as high pressure phase by a few scientists. For synthesis shock wave methods [25, 26] were used and also reactions at normal pressure with photon [27] or electron [28, 29] assistance. In a special three-dimensional phase-diagram (pressure, temperature, electrical field) the existence of the metastable E-BN was described [30].

An X-ray spectroscopic study of crystalline E-BN in combination with a detailed literature review described E-BN as an oxygen containing compound of the type $BN_{1-x}O_x$ [31].

The expression i-BN is not for a special modification of BN. It is used by some authors to describe an h-BN with boron excess and is produced by a

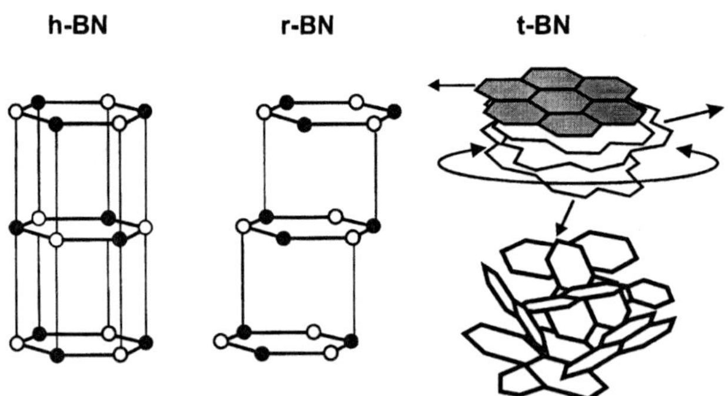

Fig. 4. Crystal structures of h-BN, r-BN, and t-BN

method where the activation energy is supplied by energetic ions [32, 33]. A broad IR peak at 1400 cm^{-1} is typically for i-BN. The hardness of i-BN is given as 26.9 Gpa [33]. Also, boron nitride layers synthesized by ion-bombardment are called i-BN [e.g. 34–36].

2.2
Conversion of c-BN into h-BN

2.2.1
Reaction at Standard Pressure

The transformation of c-BN into h-BN was investigated during DTA-analysis [22]. The SEM images show formation of differently textured h-BN at the surface of the c-BN crystallites, which indicates that there is no uniform mechanism of phase transition (Fig. 5).

The DTA measurements exhibit the stability of c-BN at standard conditions. Influences of grain size and purity (oxide content) of the cubic boron nitride crystals on the conversion temperature become obvious. Fine grained samples containing boron oxide show a significantly lower conversion temperature than coarse material (conversion at 900 °C for 1.5-μm c-BN containing boron oxide and 1500 °C for 600-μm pure c-BN).

This investigation also showed that two different routes are possible for the c-BN → h-BN transformation at normal pressure and elevated temperatures:

– Phase conversion of the solid bulk material, leading to layered textures of h-BN on c-BN crystals.

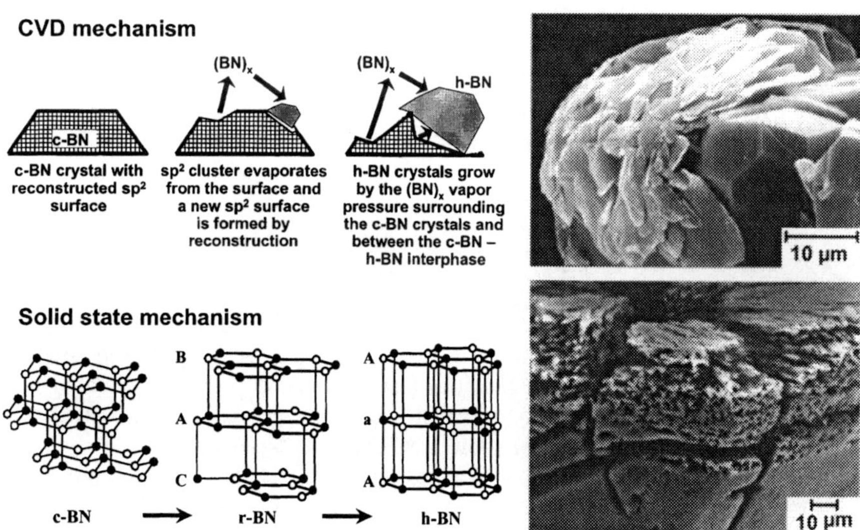

Fig. 5. Mechanisms of the c-BN to h-BN conversion (CVD mechanism and solid state mechanism) [22]

- A conversion via gas phase transport. The BN vapor pressure raises with temperature and beyond 1000 °C evaporation of $(BN)_x$ species and deposition of crystalline h-BN and BN whiskers occurs.

2.2.1.1
Solid State Mechanism

The atomic layers from c-BN (ABCABC) have to rearrange into an ABAB stacking sequence of h-BN during the solid state phase conversion. A possible mechanism would be the intermediate formation of the rhombohedral BN phase (r-BN) with ABCABC stacking. The r-BN phase is structurally related to the hexagonal phase, but only differs in the d-values (h-BN: d = 6.66 Å; r-BN d = 10.0 Å) of the layers (Fig. 5b). Subsequently the rhombohedral phase is transformed into the hexagonal modification at the reaction temperatures [10].

2.2.1.2
Gas Phase Mechanism

For gas phase reactions volatile compounds are necessary. The vapor pressure of h-BN (1 mbar at 1300 °C, 14 mbar at 1600 °C [37, 38]) is high enough for the reaction. The evaporation from the c-BN surface can be explained by formation of distinct sp^2-bound BN regions (atomic layers), which may lead to sublimation of $(BN)_x$ species in various degrees of aggregation. Theoretical investigations show that the c-BN crystals consist of a sp^2 terminated surface which reveals the CVD mechanism [39]. Grain size, crystallinity, and impurities influence the BN evaporation and thus the wide temperature range for the phase conversion, (more than 1100 K measured by DTA experiments [22]) can be explained. At high heating rates the onset of the conversion temperature seems to be higher, because the phase transition is hindered by kinetic activation.

As demonstrated in SEM pictures (Fig. 5), the phase conversion starts at the surface of the c-BN crystals, which leads to the formation of h-BN platelets and whiskers (Fig. 5). This clearly shows that a CVD mechanism is involved and the solid state mechanism is not the only reaction pathway [22].

2.2.2
Transformation at Increased Pressure

The transformation of h-BN into c-BN (at 6.5 GPa) and the reverse transformation of c-BN to h-BN (from 0.6 to 2.1 GPa) were investigated in a Li_3N-BN catalyst system [40]. Synchrotron radiation was used to check the phases and to examine reactions between the BN-phases and the catalyst.

The reverse transformation was investigated in a temperature range from 800 to 1200 °C. The reaction started at the same time at which a melt was formed. The experiments described showed that the transformation from c-BN to h-BN took place within minutes. These experiments revealed that c-BN is the stable phase at standard conditions in the presence of a Li-catalyst [41].

2.3
The BN Phase Diagram

In 1963 a phase diagram was established by Bundy and Wentorf [19], based on data of Wentorf [42] and experiments carried out at pressures higher than 4 GPa. This phase diagram described c-BN as the stable phase at standard temperature and pressure (Fig. 6). In 1975 a new phase diagram was published by Corrigan and Bundy [11], showing the c-BN/h-BN equilibrium line similar to the graphite/diamond line in the carbon system.

The phase diagram of Corrigan and Bundy had been considered to be correct until 1987 when Leonidov et al. [43] published fluoro-calorimetric results for burning c-BN, and additional calculations from Solozhenko and Leonidov [44] followed in 1988. These papers described c-BN as the stable phase – up to 1300 °C. A comparison of the data for burning c-BN and h-BN confirm the c-BN stability:

$$\text{h-BN} \, \Delta H^{\circ}_{r\,(298.15\,K)} = -884.91 \pm 1.15 \, \text{kJ/mol} \quad [45]$$

$$\text{c-BN} \, \Delta H^{\circ}_{r\,(298.15\,K)} = -869.2 \pm 2.0 \, \text{kJ/mol} \quad [43]$$

Further results were reported by Maki et al. in 1991 [46] and by Solozhenko in 1993 [47]. However, the thermodynamic data show large discrepancies (Table 2 [44, 47], Table 3 [43, 45, 46, 48–50]) and therefore the differences in the phase diagrams are easy to explain.

It can be summarized that c-BN is the stable phase at room temperature but there are still discrepancies about the phase transformation lines and thermodynamic data. Several review articles draw conclusions from the available results (Fig. 7) [11, 41, 51–54] but there are also new results describing c-BN as metastable at standard conditions [55].

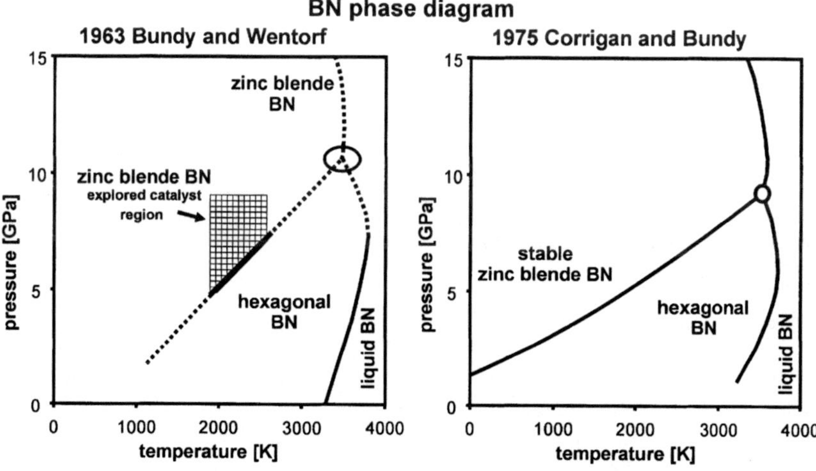

Fig. 6. BN phase diagrams described by Bundy and Wentorf [19] and by Corrigan and Bundy [11]

Table 2. Thermodynamic data for the conversion of h-BN into c-BN at 298 K (25 °C)

	[44]	[47]
Enthalpy ΔH_r° (298.15 K) [kJ/mol]	−16.3 ± 2.7	−16.5
Entropy ΔS_r° (298.15 K) [J/mol · K]	−8.22 ± 0.01	−8.56
Free enthalpy ΔG_r° (298.15 K) [kJ/mol]	−13.9 ± 0.17	−3.6

Table 3. Standard free energy of formation for h-BN and c-BN at 298 K (25 °C)

$H_f^\circ{}_{298\ K}$ [kJ/mol]

Reference	[48]	[45]	[49]	[46]	[50]
h-BN	−250.914 ± 1.548	−251.0 ± 1.5	−252.8	−254.0	−254.4
Reference	[43]		[46]		
c-BN	−266.8 ± 2.2		−266.1		

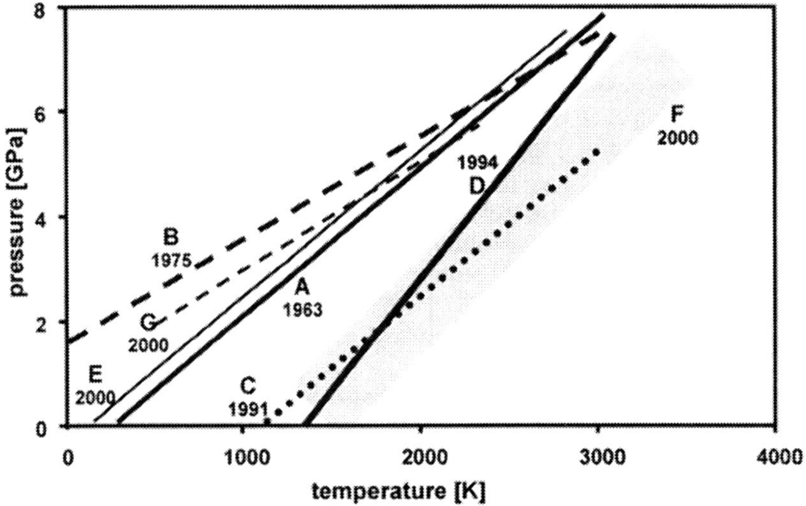

Fig. 7. Summary of published results about the location of the c-BN/h-BN phase boundary. A – Bundy and Wentorf (1963) [19]; B – Corrigan and Bundy (1975) [11]; C – Maki et al. (1991) [46]; D – Solozhenko (1994) [51]; E – Will et al. (2000) [41]; F – Will et al. (2000) calculated from Solozhenko's data [41]; G – Fukunage (2000) [55]

2.4
Characterization of BN Products

Due to the complexity of BN-structures and atomic bonding situations, the characterization of BN-phases by spectroscopic methods (e.g., IR and Raman) is difficult. It is not possible to identify BN phases using only one analytical method. For example, the X-ray diffraction peaks of c-BN correspond to those of Cu, Ni, and many other cubic phases. Elemental composition must be known or measured to be sure that no other phases are present.

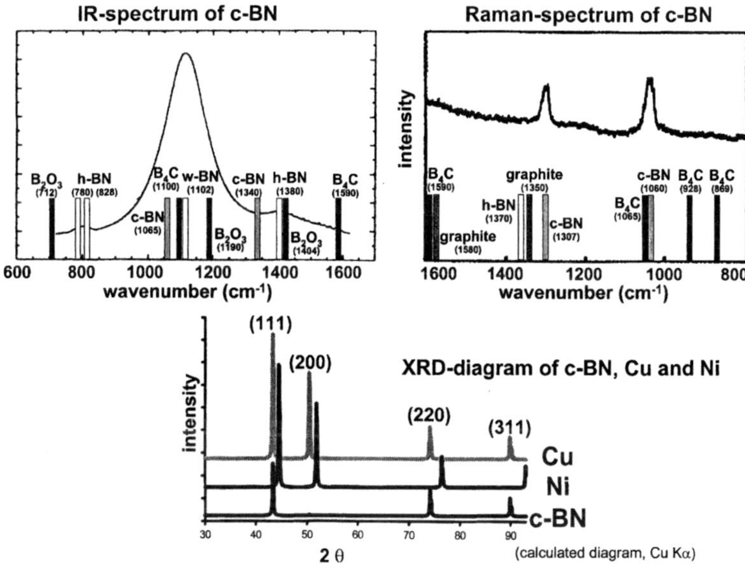

Fig. 8. IR- and Raman spectra and X-ray diffractogram of c-BN. Peak positions of other compounds in the relevant region are marked

Some data for analytical characterization are summarized in the following sections (Fig. 8).

Infrared spectroscopy (IR) or Fourier transform infrared spectroscopy (FTIR) are often used to characterize BN products. If pure BN mixtures with B:N ratio of 1:1 are analyzed, it will be easy to distinguish between h-BN and c-BN. However, if the chemical composition of the sample is unknown, many artifacts can occur and a clear statement is often not possible.

To identify c-BN, the characteristic transverse optical mode (TO) at 1065 cm^{-1} and longitudinal optical mode (LO) at 1340 cm^{-1} have been described [56]. When investigating commercial c-BN, commonly only one IR-peak between 1050 and 1100 cm^{-1} is observed.

The situation is more complex for h-BN, t-BN, and a-BN, because all of them show peaks between 780 and 1370 cm^{-1}, which makes it impossible to distinguish among these phases.

Table 4 summarizes spectroscopic data for BN phases prepared by different methods [36, 56–65].

Raman spectroscopy is useful to distinguish between h-BN and c-BN, too. As described above, impurities as well as non-stoichiometric mixtures can result in misinterpretations.

In the case of c-BN, two characteristic peaks are observed, and only one peak for h-BN.

The two peaks for c-BN are described by several authors as a TO-mode at 1055–1057 cm^{-1} and an LO-mode at 1305–1306 cm^{-1} respectively [66–68]. The crystallinity seems to be important for the Raman peaks of c-BN, because

Table 4. IR-peaks of c-BN and h-BN samples synthesized by different methods

c-BN	h-BN		Reference	Method
	TO	LO		
–	810	1310	[57]	CVD
1065			[56]	HP-HT synthesis
1050	800	1400	[58]	Laser PVD
1050	800	1390	[36]	Ion-plating
1060	780–800	1380	[59]	ECR PACVD
1060	800	1370	[60]	RF sputtering
1060–1080	780	1390	[61]	PACVD
1065	770	1380	[62]	Sputtering
1100	770	1370	[63]	Pulsed laser deposition
1100	800	1380	[64]	PACVD
1110	800	1364	[65]	ECR PACVD

in the case of very fine grained c-BN (found mainly in PVD layers) these peaks have not been observed [36].

For h-BN the characteristic peak is located at 1367 cm^{-1} [67].

X-ray diffraction is another possibility to distinguish between c-BN and h-BN. In that case again the elemental composition of the sample is important, because the diffractograms of many cubic substances can mimic the one of c-BN (e.g. Cu, Ni, etc.). The peak positions and their intensities are mainly influenced by grain size, stress in the layers and impurities.

Summing up, it can be said that for the characterization of c-BN in unknown samples (mainly PVD and Plasma-CVD deposits), the results of only one analytical method are not sufficient for a definitive characterization. Measurements of the elemental composition in combination with IR and/or Raman and/or X-ray diffraction are necessary to ensure that c-BN is present.

3
Hexagonal Boron Nitride (h-BN)

Hexagonal BN is a high-temperature solid lubricant, good thermal conductor, and good electric insulator. The specific gravity is low, it is stable in air up to 1000 °C, under vacuum up to 1400 °C, and in inert atmosphere it can be used up to 2800 °C. Hence, the maximum application temperature is higher than that of Si_3N_4, Al_2O_3, or SiC. The temperature resistance of BN can be compared to MgO, ZrO_2, or CaO but BN shows higher thermal shock resistance than these oxides. BN is chemically inert and not wetted by many metallic (Al, Cu, Zn, Fe, steel, Ge) and non-metallic (Si, B, glass, cryolite, halides) melts. The hardness is similar to graphite and therefore BN materials, produced by hot-pressing, can be machined easily for low costs. Close tolerances for h-BN components can be reached using conventional shop tools. The thermal and mechanical properties show an anisotropic behavior parallel and perpendicular to the hot-press direction. Detailed information about the chemical and physical properties of h-BN is given in [69, 70].

This combination of excellent properties of hexagonal BN (h-BN) opens a huge range of technical applications.

3.1
Synthesis Methods

Balmain first synthesized BN in 1842 by reaction of molten boric acid with potassium cyanide [1]. More than 100 years later the commercial production of BN was established. Although there are a lot of other general methods for producing BN, principally two reactions are used on the industrial scale.

3.1.1
Boric Acid with Carrier Substances

Boric acid with ammonia reacts in the presence of carrier substances $(Ca_3(PO_4)_2, CaCO_3, CaO, BN, Zn\text{-borate})$ [71–75]. The carrier substances prevent the formation of a homogeneous melt of boric acid, which is not suitable because of its minimal surface: At reaction temperatures exceeding 700 °C a thin film of molten boric acid covers each carrier substance particle. Because of the large surface a full reaction of the boric compound with ammonia is possible. After the reaction the carrier is leached with HCl and the remaining h-BN is washed with water. A second reaction at temperatures exceeding 1500 °C with ammonia follows, resulting in h-BN powders with 97% purity. The h-BN crystallites are thin hexagonal platelets with a thickness of about 0.1–0.5 µm and a diameter up to 5 µm.

3.1.2
Boric Acid with Organic Nitrogen Compounds

The second important way to produce h-BN is the reaction of boric acid or alkali-borates with organic nitrogen compounds (melamin, urea, dicyanamide, guanidine) in nitrogen atmosphere [76–80]. These reactions are carried out at temperatures between 1000 °C and 2100 °C in N_2 atmosphere. Before final thermal treatment, the product can be washed with methanol or diluted acids in order to remove all non-reacted products. For removing oxygen impurities a thermal treatment at 1500 °C in inert N_2 or Ar atmosphere is used. BN with turbostratic structure (t-BN) is obtained which is characterized by partial or complete absence of three-dimensional order in the stacking of its atomic planes [81].

Annual world production of h-BN powder is approximately 400 tonnes at a price of $50–150 per kg.

3.1.3
Various BN Synthesis Methods

– In carbothermal reactions BN is synthesized by reduction of boric acid or borates in nitrogen atmosphere at 1000–1500 °C. One example is the reduction of boric acid or borates with carbon. The precursors are mixed intensively and after drying, the mixture is heated up to 1200–1500 °C in nitrogen atmosphere, remaining at maximum temperature for 60–600 min. To enhance the yield of BN, catalysts ($CaCO_3$, MnO_2) are sometimes added to the mixture. However, too high carbon content of the mixture leads to the formation of boron carbide [82–85].

- Reaction of amides, cyanides, cyanamides, and thiocyanates with boron compounds. $NaNH_2$ forms boron nitride with boric acid or sodium borates at 300–500 °C in ammonia atmosphere. At the end of the reaction the temperature is increased to 1000 °C for the decomposition of the remaining nitrogenous compounds. After washing with water, the BN powder is stabilized by annealing at 1800 °C in inert atmosphere [86]. Cyanides (NaCN, KCN) react with boric acid at 800–1500 °C by forming BN. As a by-product, CO gas and C are produced [87]. The reaction of cyanamides (CaNCN) with boric acid is carried out at 1400–1750 °C in a 93% N_2 – 7% H_2 atmosphere. After reaction the product is washed with diluted HCl. Carbon, which is produced as a by-product during the reaction, is burned in air at 1000 °C [88].
- Preparation of B-N containing precursors with subsequent decomposition is also a suitable method to produce BN. The adduct BF_3-NH_3 is formed by reaction of BF_3 with gaseous, liquid, or aqueous ammonia. From aqueous solutions the BF_3-NH_3 adduct can be precipitated by adding NaOH. After drying the precipitate it is heated up to 800 °C in inert atmosphere (N_2, H_2, or NH_3), and afterwards the BN is formed by pyrolytic reaction. After washing with hot water the h-BN is obtained [89, 90].
- Fine and ultra-fine BN is used for lubricants and toners and can be produced by combustion of boron powder at 5500 °C in a nitrogen-plasma [91–93].
- Other ways for preparing BN are the reactions of calcium-boride (CaB_6) with additions of boric acid in nitrogen atmosphere at temperatures exceeding 1500 °C [94, 95], or the synthesis from iron boride (FeB) with ammonia at 550 °C and subsequent annealing in ammonia at 1000 °C [96].
- Growing of h-BN crystals is possible in molten sodium at temperatures between 700 and 800 °C. Starting materials are boron and NaN_3 powders which are sealed in a stainless steel tube and heated up [97].

3.1.4
Gas Phase Deposition

Using thermal CVD methods with B_2H_6–NH_3–H_2 [98, 99] or BCl_3–NH_3–H_2 [100] gas mixtures, different BN-layers can be deposited; e.g., h-BN, t-BN, or a-BN. BN with higher boron contents can be deposited at enhanced deposition temperatures. To deposit crystalline h-BN from the gas phase, temperatures above 1100 °C and a N/B ratio of 10:1 are necessary.

In the last century a dramatic increase in the number of reports and patents describing the deposition of h-BN has taken place. Different methods for coating on a wide range of materials have been developed. The activation of the gas mixture can be achieved by various methods, e.g., plasma-assisted CVD as well as laser ablation techniques of BN bulk materials. The plasma is generated either by microwave discharge or by an RF (radio-frequency) discharge. The CVD process can also be conducted by laser driven reactions and pulsed laser deposition. A wide range of different gaseous precursors is used in all the CVD techniques. The gas mixture may consist of different boron and nitrogen sources like B_2H_6, BCl_3, BBr_3, and NH_3, N_2, respectively [101, 102].

Most of the reports on h-BN deposition during the last years are in combination with the low-pressure nano-cBN deposition by PVD methods because a h-BN interlayer is formed before the c-BN is able to nucleate.

Layers of pure pyrolytic BN are generally produced by thermally induced CVD processes on graphite substrates. The reaction is performed using BCl_3, NH_3, and N_2 at low pressures (0.7–70 mbar) and temperatures in a range between 1500 °C and 1900 °C. For special applications, e.g., coating ceramic fibers for reinforcing purposes or nuclear fuel pellets [103], the temperature is kept below 1100 °C. Vacuum evaporation of B-trichloro-borazine and decomposition on extremely hot surfaces (e.g., graphite or tungsten) gives pyrolytic BN layers [104].

3.2
Applications of h-BN

Hexagonal boron nitride is an interesting construction material because of its high heat resistivity and stability against oxygen. Powders of h-BN are used as lubricants, for producing ceramic parts, and for coatings; all applications are suitable up to high temperatures.

3.2.1
Lubricants

Graphitic BN (h-BN) is used as lubricant with low friction in numerous applications. Compared to graphite the h-BN can be used as lubricant in an oxidizing atmosphere up to 900 °C as well as at extremely low temperatures, e.g., in space because no water inclusions between the atomic sheet layers are present (graphite always contains small amounts of water between the layers). Due to its excellent resistance against oxidation, its extremely low friction coefficient, and its chemical inertness, h-BN can be inserted into alloys or ceramics [105]. It can be used as a solid surface lubricant [106] or added to a liquid to get dispersions with lubricating properties.

3.2.1.1
Liquid Lubricants

Lubricating dispersions containing h-BN are mainly water based, oil based (mineral oil, silicone oils, highly viscose organic components), or water/oil mixtures [107, 108]. Stabilizers, thickeners, oxidation-preventing additives are added to the dispersions.

3.2.1.2
Solid Lubricants

To produce solid based lubricants, h-BN is added to plastics, rubbers, and resins in various amounts. BN dip coated surfaces are used for single-use application.

BN is used for reinforcing ceramics and alloys to reduce wear and friction and thus obtain self-lubricating parts [109–111].

Lubricant materials containing BN are used as bearing materials for high temperature applications and as sliding contact materials made of alloys or porous ceramics filled with Cu, Ag, Pb, or graphite [112–114]. BN powder serves as an additive in paints, and it acts as a lubricant agent for casting and forming processes.

BN powder is used to produce lubricants for high pressure purposes [115–117], for polishing materials, as well as for HIP (hot-isostatic pressing) techniques to prevent reactions between the powder mixture and the mold.

3.2.2
Ceramics Containing h-BN

3.2.2.1
HP-BN and HIP-BN

Dense shapes of h-BN are made exclusively by hot-pressing (HP) or hot-isostatic pressing (HIP) of BN powders using boric oxide as a sintering additive [118, 119]. The BN powders should have a fine grain size, a free boron oxide content as sintering aid, and disordered lattice or turbostratic structure. The resulting shapes are soft and can easily be machined to the desired size but the big waste of material during shaping makes the products expensive. Hot-pressed BN ceramics show a significant anisotropy in thermal expansion and thermal conductivity as well as in strength and Young's modulus, which varies with the hot-press direction. In this context BN ceramics produced by HIP technique show isostatic behavior and nearly theoretical density.

HP-BN is an excellent thermal conductor and an electrical insulator, which makes it ideal for electronic applications. As an insulator it has a dielectric constant of about 4 and a dielectric strength almost four times higher than that of alumina. As a thermal conductor BN exceeds almost all other electrical insulators while maintaining high strength and low thermal expansion. This is an ideal combination for heat sinks and substrates. Compared to other favorable materials, e.g., BeO or Al_2O_3, BN is easier to form, and a smooth surface is achieved without high cost for finishing. Pure BN ceramics are used as break rings in continuous casting of steel or in the non-iron industry, for crucibles, tubes, or plates, seal rings for gas sensors, moulds for hot-pressing of ceramics, neutron absorbers and shields for nuclear reactors and

Fig. 9. Boron nitride ceramics. BN ceramics with SiC addition for break rings in continuous ferrous metals casting. Hot-pressed and HIPed BN crucibles and electrical insulators. (Courtesy of ESK-Kempten, a company of WACKER-Chemie)

components for high-temperature electric furnaces (Fig. 9) [120]. Other applications are insulators and source holders for ion implantation systems, furnace vents, stacks and fixtures, welding tips for plasma arcs. Because of its stability at high temperatures and chemical inertness against carbon and carbon monoxide up to 1800 °C, BN is used as a refractory ceramic [121].

3.2.2.2
Nitride Ceramics with BN

Due to the good thermal conductivity and electrical insulating properties AlN ceramics are used in the electronic/electrical industry for integrated circuits and integrated circuit package materials. BN additions increase the machinability, and they enhance the resistance against erosion by molten metals [122]. The most common way to produce such composites is hot pressing or hot isostatic pressing at temperatures of 2000 °C and at pressures near 2000 bar [123]. Sintering aids, which are necessary for densification, can be B_2O_3, Y_2O_3, Al_2O_3, CaF_2, MgO, Si, Al_2O_3, CaB_6, MgB_x, and mixtures of them.

An addition of BN powder to Si_3N_4 ceramics improve the thermal shock resistance. These composites show excellent stability toward molten steel. Si_3N_4-BN composites are produced either by hot-pressing or HIP or via reaction-sintering, starting with BN powder and elemental silicon and sintering in nitrogen atmosphere at 1500 °C [124]. These composites are used to dope silicon wafers, for heat-exchangers, and for nozzles in continuous steel casting [125, 126].

3.2.2.3
Mixed Nitride – Oxide Ceramics with BN

SiAlON (Si_3N_4-Al_2O_3) ceramics reinforced with BN powder are used as break rings in continuous casting and for nozzles. The addition of BN provides better thermal shock resistance [127].

TiN-based composites with additions of Al_2O_3 and BN show excellent thermal shock resistance as well as good corrosion resistance. These composites are stable towards molten metals and are used for abrasives and cutting tools [128, 129].

3.2.2.4
Oxide Ceramics with BN

Oxide ceramics (Al_2O_3, ZrO_2) can also be reinforced with BN powder. They are often sintered by hot-pressing with sintering aids like boron oxide and calcium oxide. The resulting composites show higher thermal shock resistance and excellent corrosion resistance. Therefore they are used as material for metal casting or, generally, for materials in contact with molten metals [130]. With increasing BN content the composites show better machinability, and therefore the production costs decrease.

Borosilicate glasses, phosphate glasses, glass ceramics, and enamels are reinforced with BN powder for enhancing the strength of the composites [131, 132].

3.2.2.5
Borides and Carbides with BN

BN is used for the production of TiB_2-BN and TiB_2-AlN-BN composites. High temperature applications are evaporator crucibles and boats used for vacuum metallization of plastics, paper, textiles, and glass. When the material is soaked in molten metals it can be used for electrodes. The composites show high corrosion resistance against molten metals as well as good electrical conductivity (TiB_2 has high electrical conductivity). The crucibles/boats are heated in direct current, and by varying the BN content the electrical conductivity of the boats can be controlled (Fig. 10) [133, 134]. However, to enhance the wettability of the boat-surfaces for molten aluminum, the BN surface layer is removed by laser sputtering methods. A film of molten aluminum has lower corrosive potential than molten Al-pearls and therefore this step is necessary. Today more than 70% of the BN world production is used for producing TiB_2-BN composites.

SiC-BN composites show good thermal conductivity. The electrical resistivity depends on the BN content and can be adjusted from a few Ω/cm to more than 10^{10} Ω/cm. When the SiC content of the composites is high, the composites can be used as cutting tools [135]. BN additions reduce the friction coefficient. Thus these composites are used for sliding parts [136].

3.2.3
Coatings

Coatings of h-BN are used in a wide range of applications due to the extraordinary properties of BN. BN coatings are used in the steel and iron

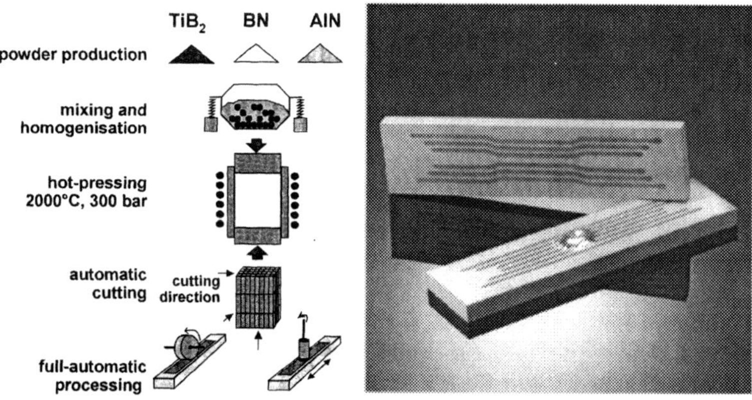

Fig. 10. Production sheet for TiB$_2$-BN-AlN evaporation boats. (Courtesy of ESK-Kempten, a company of WACKER-Chemie)

industry to enhance the corrosion resistance as well as to reduce the wear on sliding parts like the crankshafts for compressors. Because of the non-wettability of BN against many metallic and non-metallic melts [137] various parts are often coated with BN, e.g., silicon in the semiconductor industry, or BN is used for linings to prevent reactions with Si or Al. Coatings can be produced by CVD techniques to obtain well-crystallized linings with relatively high strength. Also BN layers can be achieved by spraying, brushing, dipping, or pouring [138]. The coatings provide improved high temperature electrical insulation for semiconductor industry. For mold equipment BN coatings reduce sticking and reactions between powder and mold. High temperature lubrication properties are important in steel manufacturing processes like casting. On BN coated pre-forms, metal layers can be deposited, allowing one to get the free-standing metal sheets by easily peeling-off [139].

BN coating on SiC fibers or Al$_2$O$_3$ fibers reduces the thermal mismatch between fiber and matrix and also reduces fiber-matrix interfacial shear strength in ceramic matrix composites leading to higher overall strength and toughness.

3.2.4
Pyrolytic BN

Pyrolytic BN is used in a wide range of applications. Due to its high purity, pyrolytic BN is used in equipments for semiconductor materials, e.g., for single-crystal production in the Czochralsky apparatus as well as in horizontal and vertical Bridgman apparatus. Usually the amount of impurities is low (typically a few ppm). In the case of growing GaAs single-crystals it is important to prevent silicon impurities, and therefore instead of using silica parts, pyrolytic BN coated crucibles are used [140]. Due to the non-reactivity with the melt and the non-wettability, the pyrolytic BN parts can be used several times and therefore the costs are low. In the semiconductor industry pyrolytic BN is used for crucibles and linings, covers of heating elements, in thermal evaporation of elements to coat semiconductor materials with metals, e.g., Al_2O_3, and for coating materials to perform CVD [141].

3.2.5
Electronic/Electrochemical Applications

The excellent insulating and dielectric properties of BN combined with the high thermal conductivity make this material suitable for a huge variety of applications in the electronic industry [142]. BN is used as substrate for semiconductor parts, as windows in microwave apparatus, as insulator layers for MISFET semiconductors, for optical and magneto-optical recording media, and for optical disc memories. BN is often used as a boron dopant source for semiconductors. Electrochemical applications include the use as a carrier material for catalysts in fuel cells, electrodes in molten salt fuel cells, seals in batteries, and BN coated membranes in electrolysis cells for manufacture of rare earth metals [143-145].

3.2.6
Filler Material

Additions of BN powder to epoxies, urethanes, silicones, and other polymers are ideal for potting compounds. BN increases the thermal conductivity and reduces thermal expansion and makes the composites electrically insulating while not abrading delicate electronic parts and interconnections. BN additions reduce surface and dynamic friction of rubber parts. In epoxy resins, or generally resins, it is used to adjust the electrical conductivity, dielectric loss behavior, and thermal conductivity, to create ideal thermal and electrical behavior of the materials [146].

Ultra-high purity BN powder enhances the performance of cosmetics products [147]. BN powder is added to pencil lead compositions, paints and cement in dentistry and medicine.

3.2.7
BN Fibers

BN fibers can be made by decomposition of hydrated cellulose impregnated with boric acid or ammonium tetraborate ($NH_4B_4O_7$) in ammonia and nitrogen atmosphere at elevated temperatures (>1000 °C) [148, 149].

Another method is to extrude borazine network polymers ($[-B(NH_2)-N(C_6H_5)-]_3$) with addition of B_2O_3 at T > 300 °C into fibers and subsequently firing in nitrogen stream at 1800 °C [150].

BN fibers are used for reinforcing ceramic materials (e.g., Al_2O_3, Si_3N_4, SiC) to enhance mechanical properties as well as to extend the range of possible applications. They serve as reinforcement of organic polymers (e.g., epoxides, polyether-polyketones, polyphenylensulfides) which exhibit good thermal conductivity and low thermal expansion.

4
c-BN

For industrial synthesis of c-BN only the high-pressure high-temperature (HP-HT) methods are relevant. During the last years the deposition of c-BN coatings mainly produced by PVD and Plasma-CVD methods has been investigated. Due to the difficulties of growing large c-BN crystals and depositing thick layers, the low-pressure synthesis method is not commercially used today.

4.1
High-Pressure High-Temperature Synthesis

The high-pressure high-temperature (HP-HT) synthesis was the first method [2, 42] to grow c-BN, and until now it is the only one available for industrial production. Investigations in this field have the aim to reduce pressure and temperature for the process and to find new catalysts to grow larger c-BN crystals. The fact that c-BN is a stable phase raises hopes that further decrease in pressure and temperature during synthesis is possible (Figs. 6 and 7).

4.1.1
Direct Conversion of h-BN into c-BN

Bundy and Wentorf [19] showed the direct conversion from h-BN into c-BN without any catalyst at pressures up to 18 Gpa. At temperatures below 1000 °C w-BN is formed. Only by increasing the temperature to 1730–3230 °C is c-BN formed.

Conversion parameters from h-BN to c-BN are strongly influenced by the properties of the h-BN starting material:

- Pressure and temperature can be decreased using fine-grained h-BN (6.0 Gpa and 1470–1720 K) [151].
- Small-grained h-BN with low crystallinity increases diffusion rates, and c-BN can grow more easily [152].
- By using amorphous-BN it is possible to reduce the pressure to 7.0 Gpa and the temperature to 1070 K [153]. Two types of conversion mechanisms have been described:
    a) Direct conversion from a-BN into c-BN
    b) Conversion with h-BN as intermediate phase (a-BN $\rightarrow$ h-BN $\rightarrow$ c-BN)
- With turbostratic-BN the resulting c-BN crystal size is very small, but conversion conditions of 6.0 GPa and 1250 K are possible [154].
- A further decrease in conversion temperature and pressure is possible by adding impurities. When small amounts of B_2O_3 are present in the h-BN starting material, 4.0–7.0 GPa and 1500 °C are sufficient to grow c-BN [9].

For industrial synthesis the direct conversion methods are commonly not used.

A new approach in this area is the synthesis of high purity polycrystalline c-BN sintered bodies [155]. To get a full conversion of h-BN into pure c-BN, pressure and temperature must be relatively high and pure h-BN (<0.03 wt% B_2O_3) is necessary. Typical conversion parameters are 7.7 GPa, 1900–2700 °C, and 15 min reaction time. Above 2400 °C no h-BN was observed by X-ray diffraction in the sintered bodies. The c-BN grain size increases from 0.5 µm at 2300 °C to 5 µm at 2700 °C [155].

4.1.2
Catalytic Conversion from h-BN into c-BN

Up to now the so-called "catalytic process" is the only way to produce c-BN on an industrial scale. However, "catalytic" is not the correct scientific term, because the activation energy for transformation is not decreased by these substances. The substances which are used have the function of a solvent, and are responsible for the formation of c-BN. This method is successful because of the different solubilities of c-BN and h-BN in the flux. The precursor substances form a eutectic melt with the h-BN [152]. If the reaction conditions are in the domain of stable c-BN, spontaneous crystallization takes place and the c-BN growth rate is relatively high.

The domain for c-BN formation and the solubility conditions are strongly influenced by the flux composition. In the catalytic process h-BN is mixed with the catalyst or just only brought into contact. In some experiments it has also been shown that c-BN can nucleate on substrates placed in the reaction chamber [156] (Fig. 11).

In the text below an overview of different solvents, and their properties, for the production of c-BN is given.

4.1.2.1
Alkaline and Alkaline Earth Elements

Lithium, magnesium, and potassium were already used by Wentorf in 1957 for the first reported synthesis of c-BN [2]. First the melted metals react with h-BN to the corresponding nitrides (Li_3N, Mg_3N_2, Ca_3N_2) and

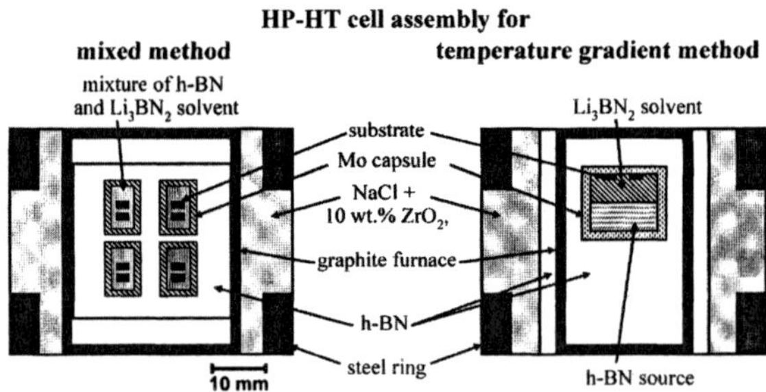

Fig. 11. High-pressure high-temperature reaction chambers for the synthesis of c-BN (substrates were added to study the c-BN nucleation) [156]

elemental boron ($3Li + BN \rightarrow Li_3N + B$). In a second step the nitrides and h-BN react to a eutectic mixture. When a condition is reached, at which c-BN is the stable phase, the rest of the h-BN can be dissolved in the flux and c-BN crystallizes. The color of the obtained powder is black or brown, which can be explained by the incorporation of excess boron in the c-BN lattice [2, 157].

4.1.2.2
Alkaline and Alkaline Earth Nitrides

Starting with nitrides of the alkaline and alkaline earth group elements no elemental boron is formed and the c-BN produced is yellow and transparent. Both the quality and the yield of c-BN are increased.

In 1961 Wentorf [157] assumed the existence of a lithiumboronitride-phase in the Li_3N-BN system, which was identified in 1972 by DeVries and Fleischer [158] as Li_3BN_2. At high pressure Li_3N and h-BN react to Li_3BN_2, forming a melt at HP-HT conditions and acting as solvent for h-BN. Because the solubility of c-BN in the melt is lower than that of h-BN, nucleation and growth of c-BN takes place. For the Li_3N-BN melt, the BN diffusion coefficient in the melt was calculated as $3 \pm 1 \times 10^{-7}$ cm^2 s^{-1}, at the synthesis conditions of 6.6 GPa and 1770 K [159]. The Li_3N-BN system is often used to study the BN phase diagram [40, 41, 160] and to produce c-BN in industrial scale.

By using Mg_3N_2 for the synthesis of c-BN a similar mechanism is observed. Endo et al. [161] found $Mg_3B_2N_4$ as active flux in the system Mg-BN. In the system $Mg_3B_2N_4$-h-BN on eutectic point exists at 1568 K and 2.5 GPa. At 5.5 GPa the eutectic temperature decreases to 1550 K [162]. In several papers the c-BN growth is described in the Mg_3N_2-BN system at various conversion parameters [163–165]. For the direct production of c-BN compacts by catalytic conversion sintering only 1 vol.% Mg_3BN_3 was used. The compacts contain 3–10 µm c-BN grains and the Mg_3BN_3 is located at the grain boundaries [166].

Similar to the magnesium boronitrides, the calcium compound Ca_3N_2 was used as catalyst by DeVries and Fleischer in 1972 [158]. Later in 1981, $Ca_3B_2N_4$ was used, which melts at 1685 K at a pressure of 2.5 GPa. $Ca_3B_2N_4$ and h-BN built a eutectic mixture at 1589 K and 2.5 GPa [167].

4.1.2.3
Alkaline and Alkaline Earth Fluoronitrides

The application of alkali and alkaline earth fluoronitrides has been investigated by Demazeau et al. [168, 169]. The synthesis of c-BN was successful by using Mg_2NF, Mg_3NF_3, and Ca_2NF under similar pressure and temperature conditions as they were found for Mg_3N_2, but the yield was much higher when using the fluoronitrides (90% instead of 30% at 7 GPa and 1470 K after 10 min).

Kinetic dependences of the conversion of h-BN into c-BN in the Li-B-N-F system have been studied by introducing about 6% additional nitrogen with addition of nitrogen compounds [170]. The stoichiometry of the c-BN was increased by adding nitrogen (synthesis pressure 4.3 GPa, 1750–1850 K).

4.1.2.4
Water and Ammonium Compounds

When water is added to the h-BN starting material, c-BN can be produced at 5 GPa and 1770 K or at 6 GPa and 870–970 K [171, 172]. These are rather low temperature and low pressure conditions compared to other syntheses. The c-BN obtained is rather fine-grained and suitable for sintering. $NH_4B_5O_8$ was identified as the flux in this process [173]. Essential for the conversion from h-BN to c-BN is the ratio NH_4^+/B_2O_3.

Unusual flux precursors like urea, ammonium nitrate, boric acid, or ammonium boron-hydride have also been studied [173]. In all these cases derivatives of amino borate form the liquid phase.

The limits for the synthesis parameters are: lower temperature limit 1070–1270 K; lower pressure limit 4.8 GPa for H_2O/4.3 GPa for urea/4.6 GPa for boric acid.

A rather interesting discovery was made by Kobayashi et al. [174]. At 1470–1870 K and 4.2 GPa, c-BN was transformed into h-BN. According to the earlier BN phase diagrams this should not have happened, because these conditions are located in the region where c-BN should be stable.

Synthesis in the $B-B_2O_3$-hBN system at temperatures >2300 K and in the pressure range of 4–6.5 GPa resulted in w-BN and c-BN phases [175].

4.1.2.5
Hydrazine

Recently Demazeau reported the use of hydrazine and lithium nitride as flux, which allowed the transformation from h-BN to c-BN at a rather low pressure of 2–3 GPa and a temperature of 1000 K [176]. Similar experiments by Solozhenko et al. [177] confirmed these results but there are still some questions about measuring the process parameters. One explanation for the effects of hydrazine is that it explodes during heating the sample, which result in higher pressures and temperatures than described.

4.1.3
Temperature Gradient Method

In this high pressure process large c-BN crystals should be grown. The key to obtain larger c-BN crystals of high quality is to control the super-saturation of BN in the solvent. A temperature gradient method with exact regulation of the temperature has been developed (Fig. 11) [156, 178].

There are two possibilities for the regulation of the temperature gradient [179]. On the one hand a vertical reactor can be used, in which a temperature gradient exists. On the other hand a molybdenum disc with a hole in the middle can be put between the h-BN and the flux, and by varying the hole's diameter the temperature gradient can be regulated.

It was possible to grow c-BN single crystals up to 2 mm in size, without using seeding crystals [178]. Process parameters were 1500–1750 °C and

5.5 Gpa, with a h-BN source and Li_3BN_2 or $Ba_3B_2N_4$ as solvents. Crystal quality of the c-BN was best in the lower temperature range where c-BN nucleation is rare and crystal growth slow. The reaction time for the larger c-BN crystals was up to 80 h.

Using the temperature gradient method it was also possible to deposit c-BN on various substrates, e.g., on diamond crystals or CVD-diamond sheets [156, 180].

4.1.4
Dynamic High Pressure Conversion

For the dynamic high pressure conversion, also known as shock wave synthesis, an explosive shock wave is used for the conversion of h-BN into c-BN. High pressure and high temperature are reached for a short period of time (milliseconds). To guarantee a fast temperature decrease a metal powder, like copper, is added to h-BN in an amount of 5% [181].

This method was first applied by Du Pont to prepare ultrafine diamond powders on an industrial scale.

Today BN synthesis by shock-wave methods is mainly used to produce superhard boron-carbon-nitrogen mixtures called "heterodiamond" [182, 183].

4.2
Low-Pressure Synthesis of c-BN

Low-pressure deposition of diamond is a commonly used industrial process [184]. Because HP-HT synthesis for c-BN and diamond works under similar conditions, it was assumed that the low-pressure synthesis of c-BN should be possible analogous to that of diamond. However, there are a lot of differences between c-BN and diamond that make a low-pressure deposition of c-BN rather difficult:

- c-BN is the stable phase – diamond is metastable (at standard conditions).
- c-BN consists of two elements – diamond contains only carbon.
- h-BN selective etching is complicated – graphite can easily be etched by atomic hydrogen.
- Stabilization of the c-BN surface is difficult – the diamond surface is stabilized by hydrogen.
- Complex precursors are needed to deposit c-BN – for diamond methane is convenient.
- Analytical characterization of c-BN is complex – diamond is identified by Raman measurements.

Deposition of nano-cBN films is possible with ion-assisted CVD and ion-assisted PVD techniques. Since the c-BN crystal size in such layers is in the range of nanometers, such layers should be called nano-cBN. The amount of grain boundaries in such materials is rather high and this fact should not be negated by using the notation "pure c-BN".

4.2.1
Selective Etching of h-BN and c-BN

For a successful CVD-synthesis of c-BN a reaction system has to be found, where c-BN is deposited and h-BN and amorphous BN formation can be prevented or where the undesirable phases can be removed by selective etching during the deposition.

For the carbon system the atomic hydrogen acts as medium for selective etching and stabilization of the diamond surface.

In order to find a substance which allows selective etching of h-BN, several compounds (e.g. atomic hydrogen, chlorine, mixtures of chlorine with hydrogen, BF_3) have been tested. Of these substances BF_3 shows the best selectivity for etching h-BN compared to c-BN (Fig. 12) [185, 186]. The disadvantage of BF_3 – in contrast to atomic hydrogen in the carbon system – is the smaller dissolution rate of h-BN (about ten times lower than in the carbon/atomic hydrogen system). The consequence is that selective etching becomes relevant only at very low BN growth rates. CVD deposition experiments with BF_3 addition always resulted in h-BN or a-BN layers [185]. Last but not least, BF_3 is a very reactive and corrosive substance which also reacts with the substrate materials. This results in additional problems during deposition of coatings on conventional tools.

4.2.2
Mechanism for Ion-Assisted c-BN Deposition

In most cases nano-cBN deposition is supported by the generation of ions. Parameters for the substrate bias (ion energy, ion mass, etc. [187]) are similar to that in PVD as well as in CVD methods. Therefore a equal growth mechanism in both methods can be considered. When starting the deposition process, commonly an oriented h-BN layer is deposited on the substrate. On

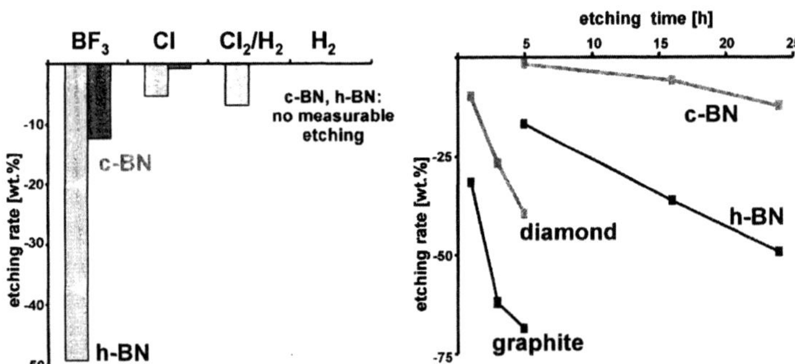

Fig. 12. Selective etching of h-BN and c-BN with various gas phases and comparison with the carbon system [185]

this interlayer the c-BN nucleates and forms a layer [188]. Because of the deposition conditions and the ion-bombardment – which is necessary for the c-BN deposition – creation of crystal defects and secondary nucleation occur, and the single-crystalline areas are very small (nm range) [189].

To describe the nano-cBN deposition four models have been proposed: the compressive stress model [190, 191], the sub-plantation model [192, 193], the selective sputter model [194], and the momentum transfer model [195].

4.2.2.1
Compressive Stress-Model

For the c-BN formation a stress threshold was observed in the deposited layers. The h-BN intermediate layer shows a preferred orientation, where the c-axis of the h-BN is parallel to the substrate. Both effects are explained by the compressive biaxial stress induced by the ion bombardment. The mechanism for the conversion of h-BN into c-BN is explained by rather high temperatures originated during thermal spikes (direct h-BN → c-BN transformation). The stress caused by the bombardment with high energetic ions is considered to be a reason for the growth of the c-BN crystals [190, 191]. A stress within the layer of up to 10 GPa has been observed. This biaxial stress causes a hydrostatic pressure up to the values usual in HP-HT synthesis.

4.2.2.2
Sub-Plantation Model

Another possibility to explain the ion assisted c-BN deposition is the sub-plantation model. The nucleation of c-BN crystals takes place under the surface of the substrate caused by sub-plantation of the ions and stress. The sub-plantation and high nucleation rates result in the nano-cBN coatings.

4.2.2.3
Sputter-Model

The essential mechanism in the sputter-model is that h-BN can be removed more easily by selective sputtering than c-BN (if the BN mixtures are deposited simultaneously and the h-BN is selectively etched, the c-BN layer remains) [187, 196].

4.2.2.4
Momentum Transfer Model

A correlation between the total momentum of impinging ions per deposited boron atom and the c-BN deposition has been observed. In this model, c-BN formation is correlated with the momentum-drive process, such as the formation of point defects in conjunction with the stress-induced phase transformation.

4.2.3
PVD Methods for Nano-cBN Deposition

The basic concept of the c-BN nucleation by ion-beam-deposition has been described by Weissmantel et al. in 1980 [197, 198]. Adhesion problems and difficulties in analytical characterization have caused large problems. A breakthrough of nano-cBN coatings and their various applications seems to be possible.

The methods mostly used for nano-cBN deposition are: ion-beam-assisted deposition (IBAD) [199]; mass selected ion beam deposition (IBD) [200]; ion plating [201]; RF- or magnetron sputtering [202] and laser deposition [203] (Fig. 13).

4.2.3.1
Ion-Beam-Assisted Deposition (IBAD)

Ion beam deposition employs an ion bombardment onto the substrate with high energetic nitrogen ions. Boron is vaporized as ions (mostly by an electron beam) and shot onto the substrate together with the nitrogen ions.

This method was used in 1983 by Satou and Fujimoto [204]. Boron was evaporated by an electron beam, and nitrogen ions (40 keV) were shot onto the substrate (Ta, NaCl). The deposited layers had a thickness of 6800 Å and the growth rate was up to 100 Å/min.

IBAD allows one to control all decisive parameters for the c-BN deposition (e.g. ion energy, ion flux, and ion/boron ratio). Typical deposition parameters are: 300–1200 eV ion energy, 0.4–0.8 mA/cm^2 ion current, 2×10^{-4} mbar pressure, 300 °C substrate temperature, ratio ions/boron atoms 0.5–3 [196, 199, 205–207].

4.2.3.2
Mass Selected Ion Beam Deposition (IBD)

The B$^+$ and N$^+$ ions are extracted from an ion source with high voltage (e.g. 30 kV), and after mass selection and deceleration (10 eV to 1 keV) the ions are

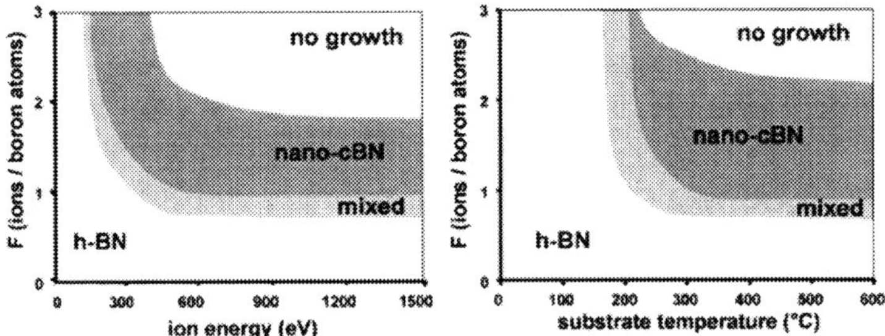

Fig. 13. Dependence of the nano-cBN deposition on deposition parameters by IBAD experiments. The results of literature data are summarized for the parameters ion energy and substrate temperature

deposited onto the substrate. Substrate temperatures between room temperature and 400 °C are possible. Ion energy, ion flux, and substrate temperature influence the quality of the nano-cBN layers [200].

4.2.3.3
Ion Plating

Ion plating and ion beam deposition differ in the use of additional ions (e.g. Ar) for deposition. The process can be divided into three steps [201]:

- Elementary boron is vaporized by an electron beam.
- Support gases (NH_3, N_2, and Ar) are activated by ionization.
- The ions are accelerated by d.c. or RF bias and deposited onto the substrate surface resulting in the growth of c-BN.

The ionization of the gas can be achieved by electron beam vaporization [208], hollow cathode discharge [209, 210], glow discharge [34, 192], arc discharge [36], or glow discharge in a parallel magnetic field [33].

4.2.3.4
Reactive Sputtering

The material is sputtered off an h-BN target and deposited on the substrate in a nitrogen/argon atmosphere. For deposition a high negative substrate bias is applied [60, 202, 211].

As sputtering sources, h-BN and B_4C compacts can be used [212, 213], and different reactive sputtering techniques can be applied by variation of the power supply on the target and the substrate (e.g. RF sputtering with h-BN targets and RF-powered substrate; RF sputtering with B_4C targets and RF- or d.c.-powered substrate; or d.c. magnetron sputtering with B_4C targets and RF-powered substrate) [214–216].

4.2.3.5
Laser Deposition

This method uses a laser for evaporating the boron compounds from a target (e.g., BN target). To regulate the flux of BN to the substrate the laser can be pulsed. Additionally, a nitrogen/argon ion-beam is generated and directed onto the substrate's surface. Various layers have been deposited, and the BN transition from substrate to the h-BN interlayer and finally to the nano-cBN has been studied in detail [203].

4.2.4
Plasma CVD Methods

Using conventional thermal CVD various BN modifications but no c-BN or nano-cBN are formed. Therefore, to synthesize c-BN a plasma is applied to activate the gas phase. Typical deposition methods are ECR plasma CVD [217, 218], ICP CVD [207], and bias enhanced plasma CVD [219].

4.2.4.1
ECR Plasma CVD

Electron cyclotron resonance (ECR) is used to deposit c-BN layers. Yokohama et al. [220] considered a negative substrate bias as an essential parameter for the deposition in ECR plasma to accelerate the ions formed in the plasma. A linear coherence was found between the bias and the etching rate of c-BN and h-BN, which was higher for h-BN. Three points are important for the crystal growth:

- The deposition of h-BN is the result of precursor substances that are created in the plasma by the reaction with electrons.
- The deposition of the c-BN phase is the result of precursor substances that are created by the reaction with Ar^+ ions.
- The growth of both phases depends on the nucleation rates, caused by the precursor substances and by the etching rate.

Typical parameters are 0.5 Pa, plasma gas Ar/N_2 mixtures, microwave power 1000 W, substrate temperature 250 °C, and boron precursor B_2H_6 [221].

4.2.4.2
ICP CVD

A very intense inductively coupled plasma (ICP) could be created, using a 13.56-MHz RF-source and coupling the power through a quartz tube into the plasma. Working pressure was 2×10^{-2} mbar and the substrate could be heated up to 800 °C [222]. The precursor was trimethylborazine $((HBN\text{-}CH_3)_3)$, which was transported in a nitrogen/argon carrier gas. Similar to other deposition processes, prior to the c-BN nucleation an oriented h-BN layer was formed [223].

4.2.4.3
Bias Enhanced Plasma CVD

Plasma CVD has been used since the middle of the 1970s. For the creation of the plasma, DC glow discharge [224], RF glow discharge [219, 225–227], microwave plasma [228, 229], or plasma jets [230] are used.

Under conditions similar to those used for the synthesis of diamond layers, only layers with a small amount of c-BN can be deposited. A lot of publications can be found about the deposition of c-BN with CVD methods, reporting different parameters that are necessary for the growth of c-BN.

Mendez et al. [231] showed in deposition experiments with RF plasma and without substrate bias that there is a coherence of the nano-cBN amount with the substrate temperature and the plasma power. Additionally a dependence on the substrate material used (e.g. Si or NaCl) could be found. Further investigations used Ar ions, where the nano-cBN amount depends on the process parameters of ion bombardment [232].

Nano-cBN in layers has been deposited by activating the gas phase with RF plasma and hot-filament. The highest amount was reached by introducing a gas mixture consisting of BH_3NH_3-H_2 into the reactor at conditions where the amount of atomic hydrogen in the gas phase reached a maximum. As a reason for the high nano-cBN amount, the selective etching of h-BN by elemental hydrogen was proposed by Saitoh and Yarbrough [229].

Matsumoto et al. [233] tried a bias assisted plasma jet with the reactive gas mixture Ar-N_2-BF_3-H_2 for selective etching of h-BN [234]. Using a d.c. bias voltage, the nano-cBN grain size in the layers could be increased from 7 nm (−150 V) to 12 nm (−80 V) [235]. By optimizing the process, nano-cBN coatings thicker than 20 μm with c-BN grain size up to 100 nm could be deposited [230].

4.2.5
Properties and Applications of Nano-cBN Coatings

One goal in the field of c-BN deposition is a coating which can be used for machining of iron-based materials. Such coatings offer the possibility to use a superhard material instead of the conventional carbide and oxide layers.

Matsumoto and co-workers showed that the deposition of thick nano-cBN layers is possible and the adhesion on Si substrates is of acceptable quality [230]. The one major question is whether this process is also suitable for hardmetal (WC-Co) substrates or not, and whether there might be any problems with the Co binder phase and the BF_3 in the gas phase?

The stress in c-BN layers caused by the high energetic ions is a problem, because many of the nano-cBN coatings delaminate during or immediately after deposition. Several investigations have showed ways to reduce the stress in the layers (e.g. buffer layers [205, 217, 236], regulation of the ion energy [207], or ion-induced stress relaxation [206]).

The outlook for industrial applications of nano-cBN coatings looks quite good, but some problems with the deposition process and the substrate materials still have to be solved.

4.2.6
The Simple Chemical Way?

Several attempts to grow c-BN at standard pressure have been performed. Similar to the high-pressure high-temperature synthesis, melts were used in the temperature range between 600 and 800 °C at standard pressure. Various compounds were mixed with fine grained c-BN powder (for seeding) and heated up to allow grain growth. The stability of c-BN in chemically active media depends on the properties of the reacting agent. Reductive media like lithium metal or lithium-boride generally lead to a strong degradation of c-BN. In this case there is no uniform reaction with c-BN, and different phases result. The reductive dissolution of c-BN first led to the formation of boron or boron-rich boron nitride $B_{50}N_2$, which further react with excess of lithium to lithium-borides.

It could also be shown that the system Li_3BN_2/c-BN interacts at elevated temperatures, leading to an obvious change in the morphology of the c-BN seed crystals. In all cases a formation of well faceted surfaces could be detected. From this system the growth of c-BN is most likely.

Actually, a spontaneous nucleation of c-BN from a degradation of Li_3BN_2 will not occur due to the fact that the less dense phase will nucleate first according to the Ostwald-Volmer rule, which is a general rule of thumb for the kinetic behavior of reactions [237]. Thus, seeding with c-BN can overcome this problem. Hence, this approach seems to be well suited to develop a low-pressure synthesis of c-BN.

4.3
Applications for c-BN Products

Because of its excellent mechanical and electrical properties c-BN is of great interest for a number of applications (e.g. grinding powders, wear parts, electronic parts, etc.). Searching the literature, the impression arises that the applications of c-BN are kept a bit secret. Most of the relevant references are patents (up to 90% depending on the topic) giving less exact data about the process. Papers published in journals giving detailed information are rare.

4.3.1
Pure Polycrystalline c-BN (PcBN)

Pure PcBN can be produced by direct conversion without catalyst [155] or with small additions of catalyst (<1 vol.% Mg_3BN_3) [166]. Compared to PcBN sintered with a binder phase, the pure PcBN materials show increased hardness (5000–5500 kg/mm^2 instead of 3000–4000 kg/mm^2). Thermal stability in air is high for the pure PcBN (1350 °C) but lower for the type containing Mg_3BN_3 (700 °C) [155]. Pure PcBN is an interesting material for wear applications and shows high thermal conductivity. Because of this instance many data are available [238–241].

4.3.2
Polycrystalline c-BN (PcBN) for Wear Applications

In the field of high temperature wear applications c-BN is superior to diamond because of its higher stability against temperature, its better oxidation resistance, and lesser reactivity with iron. The resistance of c-BN against oxidation in air up to 1200 °C is much higher than that of diamond because a protecting layer of B_2O_3 is formed by reaction with the oxygen in the air, protecting c-BN from further oxidation. Therefore, c-BN is mostly used for the processing of steel, whereas diamond is preferred for the processing of stone, ceramics, and Al-alloys. Conventional cutting and milling tools made of hardmetal, high speed steel, Al_2O_3, SiC, BC, etc. can be replaced by c-BN with all its advantages, like higher feed rate or smaller abrasion (e.g. factor of 7–50 higher compared to WC tools [152]).

The c-BN products can be divided up into dense products (e.g. for cutting, milling operations) and in porous products (e.g. for grinding).

4.3.2.1
Dense PcBN Products

A polycrystalline compact – containing c-BN powder and a second phase – can be sintered at parameters where c-BN is the stable phase (to prevent h-BN formation). Therefore, high-pressure high-temperature sintering is necessary, and the maximal diameter for the produced parts is limited by the dimension of the high pressure apparatus.

The binder phase of such materials can be a ceramic, a metal, or a resin.

4.3.2.1.1
Ceramic Binder for c-BN

Superhard BN parts are used as advanced grinding and cutting materials for processing metals. Typical trade names are Amborite, BZN HTZ, Elbor-RM, Kiborit, and Hexanit-R.

Amborite, containing AlN/AlB_2 as interphase, has a high thermal stability and is used for machining Ni-Cr containing martensitic iron alloys [242].

Elbor-RM, Kiborit and Hexanit-R contain intermetallic phases based on Al, Cr, and Zr. Additionally, Hexanit-R contains higher amounts of w-BN [243].

For the production of bodies containing c-BN and w-BN, the w-BN powder is mixed with various carbides from Group 4 or 5 (IVB, VB) elements, and afterwards the mixture is hot-pressed under high-temperature and high-pressure conditions. With this method a sintered workpiece containing w-BN and c-BN in a continuous binding phase is obtained [244].

Many different compositions of the PcBN binder phase are described in the literature (e.g. AlN [245], AlN/AlB_2 [246], Al_2O_3 [247], SiC [248, 249], Si_3N_4 [250], TiB_2, ZrB_2, HfB_2 [251, 252], TiC [253], TiN [254], borosilicate glass compositions [255]).

With a high-pressure hot-pressing method c-BN-TiC/TiN composites are prepared by sintering and subsequently heat treatment between 1000 and 1400 °C. The samples exhibit a dense polycrystalline structure of c-BN-TiN/TiC, and a thin layer of fine TiB_2 is visible at the c-BN-binder interface. Hardness decreases significantly after heat treatment [256].

A special type of such compacts is a mixture where diamond is the second phase. The products contain $\geq$90 vol.% c-BN and 2–10 vol.% diamond (particle size 0.5–2 µm). After mixing the powders, HP-HT sintering without any binder phases follows (<100 kbar and <1800 °C). The sintered products show high density [257].

4.3.2.1.2
Metallic Binders for PcBN

The bonding between c-BN and metallic binder phases is mostly quite good, due to the formation of interlayers of metal nitrides or borides. For the sintering process itself, the composition as well as the melting point of the alloys used are important. In the case of applying low sintering temperature (below the conversion temperature of c-BN into h-BN) the process can be conducted at standard pressure; when higher temperature is necessary, high-pressure must also be applied. Some examples for metallic binder are: Al, Al alloy and/or Ni, Fe, Co, Cr, and/or Mn alloys [258]; Al alloys containing Co, Cr, Fe, Mo, Ni, Si, Ti, V, W, and/or Zr [259]; Ni- or Co-based heat resistant alloys [260].

Compacts of c-BN and Al are prepared by high-pressure hot-pressing and subsequent annealing at 950 °C under pressure (3×10^{-3} Pa) for 1 h. After annealing a multilayer is formed on c-BN, starting with AlB_{10} and AlB_{12} phases, followed by a layer of columnar AlN. The thermal treatment results in an increase of mechanical strength of the sintered BN-Al system [261].

When sintering c-BN with Ti and subsequent heat treatment at 950 °C, all the Ti reacts with the BN matrix forming TiN and TiB_2 at the BN interface, and voids are observed at the BN/TiB_2 interface [262]. Similar to the Ti compacts, Zr-containing compacts are investigated in which ZrN and ZrB_2 formation is observed. Again the mechanical properties can be increased by heat treatment [263].

Another way of production is the coating of c-BN by electro-less plating with Ni-P, Ni-B, Ni-Fe-P, Ni-Cr-P, Ni-Cu-P, or Ni-W-P alloys, and mixing these powders with ≥1% of various carbides, borides, nitrides, silicides, and/or oxides. These powders are compacted and pre-sintered at 700–900 °C. Finally, hot-isostatic pressing at 1000–1400 °C and 1000–2000 bar is performed to reduce porosity [264].

In the case of a metallic binder phase (e.g. Co, Ni, Fe, and its alloys), ceramic whiskers can also be added to the c-BN [265]. The process has been described in detail for Si_3N_4 whiskers [266].

4.3.2.1.3
Hardmetal/PcBN Compacts

PcBN compacts are often bonded to a hardmetal or hard steel alloy for manufacturing a tool. This composite can be produced by direct sintering of WC-Co-PcBN mixtures or by brazing.

Onto the surface of the sintered part (e.g. hardmetal) the mixture of c-BN and/or diamond powder is bonded during high-pressure high-temperature sintering [267].

Sintered cutting tips or inserts containing c-BN can be brazed to a cemented carbide or other substrates. Brazing bond strength is increased by interlayers of various carbides, nitrides, or carbonitrides [268].

The various PcBN compacts are mainly used for machining of a wide variety of hard and/or abrasive ferrous workpiece materials up to high temperatures exceeding 1000 °C. For example: pearlitic gray cast iron; hardened ferrous metals (>45 HRc); high speed steel; case hardened steel, superalloys; sintered tungsten carbide >16% Co are machined by PcBN compacts.

4.3.2.2
Porous PcBN Products

Porous composites containing hard materials are mainly used for grinding operations. For such applications several parameters are important: e.g., the distance between the grains for good chip formation; good adhesion between the hard grains and the binder phase; good balance between the erosion of the binder phase and the hard material; porosity to allow better cooling [269].

The grinding tools can be classified by their binder phase as resin bonded, ceramic bonded, or metal bonded (sintered or electrodeposited).

Wear resistance of such tools strongly depends on the tool composition, the machined material, and the machining conditions, which makes this topic very complex [270–272].

4.3.2.2.1
Resin Bonded Grinding Tools

Grindstones are manufactured by the following steps: (1) mixing abrasive grit with epoxy resin and polymeric foaming agent, (2) casting the resulting mixture in a mould, (3) hardening the epoxy resin, and (4) heating to form designed pores which also can be formed during steps 3 and 4 [273, 274].

4.3.2.2.2
Ceramic Bonded

Grinding tools of c-BN with ceramic binders are used for grinding of hardened steel parts, e.g. roller bearing rings, tooth gears, or engine injection valves [275]. Ceramic bonded grinding wheels can be produced with porosity in the range 50–80%. Grain size of the hard material and the composition of the ceramic binder phase can be varied over a wide range [276].

Mixtures of c-BN particles with other hard materials (e.g. Al_2O_3, SiC) can be bonded by vitreous material. The binder is sintered at 850–900 °C giving a crystalline phase which increases the mechanical strength and the hardness of the tools. A typical binder consists of: SiO_2 10–20 wt%, ZnO 40–55 wt%, B_2O_3 15–30 wt%, Al_2O_3 2–10 wt%, MgO 2–10 wt%, alkali metal oxides 0.1–2 wt%, and TiO_2 0.1–2 wt% [277, 278].

A c-BN – silicon nitride ceramic composite can be produced directly by sintering a mixture of c-BN powder and Si powder in N_2 atmosphere. The composites have high resistance against heat, oxidation, and thermal shock [279].

4.3.2.2.3
Metal Bonded and Electrodeposited Grindstones

Metal bonded grindstones can be prepared either by powder metallurgy or by electrodeposition. Fabrication of metal-bonded grinding tools by green tape laser sintering is a rather new method. The porosity of Cu sintered parts is utilized for manufacturing tools containing various hard materials (e.g. c-BN or Al_2O_3) [280].

An advantage of metal-bonded grindstone (e.g. brass-bond cubic BN) is the possibility of electrolytic dressing (surface preparation). By applying voltage between the grindstone and an opposite electrode the metal bond can be selectively etched [281].

To produce grinding wheels, a foamed substrate (e.g. polyurethane foam) is coated with an electrically conducting material (e.g. graphite) and then coated with a metal layer (e.g. Ni [282], Cu, Cr, Cd, Sn [283]) by electrodeposition. Finally, a coating with abrasive particles (e.g. c-BN, diamond) is deposited to produce the grinding wheels [284]. The electrodeposition can be performed by dipping the substrates in a plating solution containing the ultrafine super-abrasives. The superabrasive surface is treated to generate a difference between its isoelectric point and the pH of the plating solution. Uniform and dense electrodeposited layers are formed [285].

4.3.3
Electronic Applications

Because of its large band gap c-BN is a good electrical insulator. Due to its extraordinarily high thermal conductivity it is used as heat sink for semiconductor laser or microwave applications.

Semiconducting c-BN was synthesized by Wentorf [286] using HP-HT methods. For p-material Be doping was used, the n-type was produced by doping the material with S, Si, or KCN.

With HP-HT methods the production of p-n modules, diodes, or LEDs [287, 288] is possible. Diodes of c-BN can be used up to 600 °C because of the large band gap and its heat resistance. LEDs emitting light from red to ultra violet can be produced. Such electronic applications are at the beginning of their development. The main problem today is the fabrication of single crystals of a sufficient size, which is not even possible with HP-HT methods. Thus, an industrial application does not exist up to now.

Sintered heat sink materials were produced of c-BN and AlN ($\leq$1% O) with a composition between 40–90 vol.% c-BN and the balance AlN. Sintering conditions were temperatures exceeding 1200 °C and pressures exceeding 40 kbar. The products have a thermal conductivity exceeding 0.6 cal/cm.s.K at 25 °C. A mixture consisting of 80 vol.% c-BN (grain size 15 μm) and 20 vol.% AlN (5 μm) gave a thermal conductivity of 1.7 cal/cm.s.K at 25 °C [239, 289].

5
References

1. Balmain WH (1842) J Prakt Chem 27: 422
2. Wentorf RH Jr (1957) J Chem Phys 26: 956
3. Mishima O, Era K (2000) In: Kumashiro Y (ed) Electric Refractory Materials. Marcel Dekker, New York Basel, pp 495–549

4. Pierson HO (1975) J Compos Mater 9: 228–240
5. Landolt – Börnstein (1978) Serie III, Band 7, Springer, Berlin Heidelberg, New York
6. Chopra KL, Agarwal V, Vankar VD, Deshpandey CV, Bunshaw RF (1984) Thin Solid Films 307–312
7. Wentzcovitch R, Chang KJ, Cohen ML (1986) Phys Rev B34(2): 1071–1079
8. DeVries RC (1972) Cubic boron nitride: handbook of properties. General Electric Company
9. Soma T, Sawaoka A, Saitoh S (1974) Mater Res Bull 9: 755
10. Gmelin (1974) Handbuch der anorganischen Chemie, Band 13: Borverbindungen, Springer, Berlin Heidelberg, New York
11. Corrigan FR, Bundy FP (1975) J Chem Phys 63: 3812–3820
12. Demazeau G (1993) Diamond Related Mater 2: 197–200
13. Taylor KM (1955) Ind Eng Chem 47: 2506–9
14. Sokolowski M, Sokolowska A, Rusek A, Romanowski Z, Gokieli B, Gajewska M (1981) J Cryst Growth 52: 165–167
15. Karim MZ, Cameron DC, Hashmi MSJ (1993) Surf Coat Technol 60: 502–505
16. Chrenko RM (1974) Solid State Commun 14: 511–515
17. Sokolowski M, Sokolowska A, Michalski A, Romanowski Z, Rusek-Manzurek A, Wronikowski M (1981) Thin Solid Films 80: 249–254
18. Hibbs LE Jr, Wentorf RH Jr (1974) Eighth Plansee Seminar 2–42, p 5
19. Bundy FP, Wentorf RH (1963) J Chem Phys 38: 1144–1149
20. Xu YN, Ching WY (1993) Phys Rev B48: 4335–4351
21. Kurdymov AV, Solozhenko VL, Zelyavsky WB, Petrusha IA (1993) J Phys Chem Solids 54: 1051–1053
22. Sachdev H, Haubner R, Nöth H, Lux B (1997) Diamond Related Mater 6: 286–292
23. Hamilton EJM, Dolan SE, Mann CM, Colijn HO, McDonald CA, Shore SG (1993) Science 260: 659–661
24. Schmolla W, Hartnagel HL (1983) Solid State Electron 26: 931–939
25. Batzanov SS, Blokhina GE, Deribas AA (1965) J Struct Chem 6: 209
26. Akashi T, Sawaoka A, Saitoh S, Arahi H (1976) Jpn J Appl Phys 15: 891
27. Sokolowska A, Olszyna A (1992) J Cryst Growth 121: 733–736
28. Sokolowska A, Olszyna A (1992) J Cryst Growth 116: 507–710
29. Michalski A, Olszyna A (1993) Surf Coat Technol 60: 498–501
30. Sokolowska A, Wronikowski M (1986) J Cryst Growth 76: 511–513
31. Batsanov SS, Kopaneva LJ, Lazareva EV, Kulikova IM, Barinsky RL (1993) Propel Explos Pyrotech 18: 352–355
32. Weissmantel C, Bewilogua K, Breuer K, Dietrich D, Ebersbach U, Erler HJ, Rau B, Reisse G (1982) Thin Solid Films 91: 31
33. Murakawa M, Watanabe S (1990) Surf Coat Technol 43/44: 128–136
34. Rother B, Zscheile HD, Weissmantel C, Heiser C, Holzhüter G, Leonhardt G, Reich P (1986) Thin Solid Films 142 83–99
35. Halverson W, Quinto DT (1985) J Vac Sci Technol A3: 2141–2146
36. Ikeda T, Kawate Y, Hirai Y (1990) J Vac Sci Technol A8: 3168–3174
37. Lorenz R, Woolcock J (1928) Z Anorg Chemie 176: 283–304
38. Campbell J (1949) J Electrochem Soc 96: 318–333
39. Widany J, Frauenheim T, Lambrecht WRL (1996) J Mater Chem 6: 899
40. von der Gönna J, Meurer HJ, Nover G, Peun T, Schönbohm D, Will G (1998) Mater Lett 33: 321–326
41. Will G, Nover G, von der Gönna S (2000) J Solid State Chem 154: 280–285
42. Wentorf RH (1959) J Phys Chem 63: 1934–1941
43. Leonidov VY, Timofeev IV, Solozhenko VL, Rodionov IV (1987) Russ J Phys Chem 61: 1503–1504
44. Solozhenko VL, Leonidov VY (1988) Russ J Phys Chem 62: 1646–1647
45. Wise SS, Margrave JL, Feder HM, Hubbard WN (1966) J Phys Chem 70: 7–10
46. Maki J, Ikawa H, Fukunaga O (1991) New Diamond Sci Technol 1051–1055

47. Solozhenko VL (1993) Thermochim Acta 218: 221–227
48. Svensk EnergiData Agersta (1986) Datenbank EKVICALC 1.21, Svensk EnergiData Agersta, Bälinge, Sweden
49. Phillips JC, Van Vechten JA (1970) Phys Rev B2: 2147–2160
50. Lide DR (1991/2) CRC handbook of chemistry and physics, 72nd edn, CRC Press, pp 5–21
51. Solozhenko VL (1995) High Pressure Res 13: 199
52. Solozhenko VL (1995) J Hard Mater 6: 51–65
53. Solozhenko VL, Turkevich VZ, Holzapfel WB (1999) J Phys Chem B 103: 2903–2905
54. Will G, Nover G, von der Gönna J (1998) SoRev High Pressure Sci Technol 7: 975–979
55. Fukunaga O (2000) Diamond Related Mater 9: 7–12
56. Gielisse PJ, Mitra SS, Plendl JJN, Griffis RD, Mansur LC, Marshall R, Pascoe EA (1967) Phys Rev 155: 1039–1046
57. Li PC, Lepie MP (1965) Am Ceram Soc 48: 277–278
58. Mineta S, Kohata M, Yasunaga N, Kikuta Y (1990) Thin Solid Films 189: 125–138
59. Weber A, Bringmann U, Nikulski R, Klages CP (1993) Surf Coat Technol 60: 493–497
60. Mieno M, Yoshida T (1990) Jpn J Appl Phys 29: L1175–L1177
61. Dworschak W, Jung K, Ehrhardt H (1994) Diamond Related Mater 3: 337–340
62. Kidner S, Taylor CA II, Clarke R (1994) Appl Phys Lett 64: 1859–1861
63. Doll GL, Sell JA, Taylor CA II, Clarke R (1991) Applications of diamond films and related materials. Mater Sci Monogr 73: 653–660
64. Becht JGM, Bath A, Hengst E, van der Put PJ, Schoonman J (1991) J Phys (Paris) C2: 617–624
65. Shapoval SY, Petrashov VT, Popov OA, Westner AO, Joder MD Jr (1990) CKC Lok Appl Phys Lett 57: 1885–1886
66. Brafman O, Lengyel G, Mitra SS, Gielisse PJ, Plendl JN, Mansur LC (1968) Solid State Commun 6: 523–526
67. Huong PV (1991) Diamond Related Mater 1: 33–41
68. Herchen H, Cappelli MA (1993) Phys Rev B47: 14,193–14,199
69. Gmelin (1991) Handbook of inorganic and organometallic chemistry, 8th edn, Boron compounds, 4th suppl, Vol 3a, Springer Berlin, Heidelberg, New York
70. Schwetz KA, Lipp A (1979) Ber Dt Keram Ges 56: 1
71. Moeser L, Eidmann W (1902) Ber Dt Chem Ges 35: 535
72. Taylor KM (1959) US Pat 2 888 325
73. Ingles TA, Popper P (1960) The preparation and properties of boron nitride. In: Popper P (ed) Special Ceramics. Heywood and Comp, London, p 144
74. Lipp A (1964) (1971)DP 1 153 DOS 1 792 741
75. Yokoi K, Ito K, Tokuda M (1995) Mat Transact JIM 36: 645
76. King EM (1967) Canadian Pat 675,630
77. Babl A, Geng H (1977) DP 1 667 538
78. Hagio T, Nonaka K, Sato T (1997) J Mat Sci Lett 16: 795
79. Chakrabartty S, Kumar S (1995) Trans Indian Ceram Soc 54: 48
80. Kawasaki T, Kuroda Y, Nishikawa H, Hara H (1997) DE 97 19,701,771
81. Kawasaki T, Nagahama K, Kuroda Y (1996) Jpn Kokai Tokkyo Koho JP 95–22830
82. Samsonov GV (1960) USSR-Pat 129 647
83. Slepstov VM, Samsonov GV (1960) Bornitrid Zhur Prikl Khim 34: 501
84. Wood AAR, Shears E C (1959) DAS 1 193 821
85. Su JY, Jha A (1996) J Mat Sci 31: 2265
86. Tagawa H, Itouji O (1962) Bull Chem Soc Japan 35: 1536
87. Conant LA, Hittle EF (1958) US Pat 2 834 650
88. Luberoff BJ (1966) US Pat 3 261 666
89. Mexer F, Zappner R (1921) Ber Dt Keram Ges 54: 560
90. Romanov VD, Samsonov GV, Nikitin DI (1959) USSR-Pat 120 509
91. Komatsu W, Morikawa T, Sedaka R (1988) Jpn Kokai Tokkyo Koho JP 63–62544
92. Canon KK Japan (1985) Jpn Kokai Tokkyo Koho JP 60–00826

93. Chen GQ, He XD, Han JC, Wood JV (2000) J Mat Sci Let 19: 81
94. Schwetz KA, Vogt G, Lipp A (1978) DOS 2 461 821 and US Pat 4 107 276
95. Zhang YM, Wang H, He XD, Han J, Du S (2001) Trans Nonferrous Met Soc China 11: 76
96. Warner TE, Fray DJ (2000) J Mater Sci 35: 5341
97. Yano M, Yap Y, Okamoto M, Onda M, Yoshimura M, Mori Y, Sasaki T (2000) Jpn J Appl Phys 39: 300–302
98. Hirayama M, Shohno K (1975) J Electrochem Soc 122: 1671–1676
99. Adams AC, Capio CD (1980) J Electrochem Soc 127: 399–405
100. Gafri O, Grill A, Itzhak D, Inspector A, Avni R (1980) Thin Solid Films 72: 523–527
101. Choi BJ, Park DW, Kim DR (1995) J Mater Sci Lett 14: 452
102. David P, Mathiot A, Lulewicz JD, Narcy B (1994) Mater Res Soc Symp Proc 327: 233
103. Durmazuçar H, Gündüz G (2000) J Nucl Mater 282: 239
104. Francis R, Flint EP, Little AD (1961) WAL-766 41: 28
105. Akaishi M, Yamaoka N, Ueda F, Sasano M (1992) Jpn Kokai Tokkyo Koho JP 90-216,361
106. Torre M (1971) Swiss Pat Ch 19,700,227
107. Matje P, Reck F, Roehlinger HU (1995) DE 94-4,405,864
108. Faustinus F, Tani M (1998) Jpn Kokai Tokkyo Koho JP 96-301,051
109. Saito T, Honda F (2000) Wear 244: 132
110. Saito T, Honda F (2000) Wear 237: 253
111. Saito T, Imada Y, Honda F (1999) Wear 236: 153
112. Takahashi K, Georuku U (1995) Jpn Kokai Tokkyo Koho JP 93-170,390
113. Haraguchi F, Kanayama H, Kamiya S, Kawakami S, Michioka H, Fuwa Y (1997) Jpn Kokai Tokkyo Koho JP 95-237,018
114. Ozaki K, Yamamoto K, Shibayama T (1997) DE 97-19,708,197
115. Ditter J, Allen R, Dorchester T, Harold T, Gerstein M, Christian JB (1967) Lubr Eng 23: 330
116. Kaldonski T (1997) Tribologia 28: 647
117. Kinoshita H, Nomura S, Mishima M (1992) EP 508,115
118. Hubacek M, Rehak B, Prnka T (1990) Hot pressing of hexagonal boron nitride. In: Exner HE, Schumacher V (eds) Adv Mater Processes Proc Eur Conf 1st, p 653
119. Hara H, Ootsu K, Yoshino N, Nakamura Y (1997) Japan Jpn Kokai Tokkyo Koho JP 09,052,771
120. Schwetz KA (1999) Nichtoxidische Werkstoffe In: Taschenbuch Feuerfeste Werkstoffe Vulkan Verlag Essen, p 130
121. Fister D (1985) Ceram Eng Sci Proc 6: 1305
122. Kurita S, Nakashima M (1999) Yoyuen oyobi Koon Kagaku 42: 99
123. Kanai T, Tanemoto H, Kubo H (1990) Jpn Kokai Tokkyo Koho JP 89-8548
124. Kusunose T, Choa Y-H, Sekino T, Niihara K (1999) Key Eng Mater 161/163: 475
125. Fukuda M, Sato Y, Ueki M (1993) Jpn Kokai Tokkyo Koho JP 92-137,596
126. Herrmann M, Schubert C (1996) DE 94-4,435,182
127. Takenouchi T, Sasaki H, Niihara K, Kususe H (1999) Jpn Kokai Tokkyo Koho JP 97-315,006
128. Rolander U, Weinl G (2000) EP 1,043,410
129. Kida O (1989) Jpn Kokai Tokkyo Koho JP 63-270,359
130. Sindlhauser P (1993) DE 91-4,143,344
131. Endo M, Ishikawa S (1997) Jpn Kokai Tokkyo Koho JP 09,278,457
132. Unuma H, Kikuchi A, Nakagawa T, Yamamoto T (1994) Hokkaido Kogyo Gijutsu Kenkyusho Hokoku 60: 1
133. Lipp A (1969) DAS 1 289 712
134. Fubacheku M, Ueki M, Sakamoto K, Sato Y (1998) Jpn Kokai Tokkyo Koho JP 97-103 283
135. Yao X, Rai G (1995) GB 94-2514
136. Moessner B, Knoch H (1996) EP 95-116,831
137. Izumitani T, Shoda K, Nakyjima M, Kurita S (1993) Jpn Kokai Tokkyo Koho JP 92-12,958

138. Advanced Ceramic Corporation (2000) Product information
139. Iwamoto Y, Ebe A, Nishama S, Ogata K (1996) Jpn Kokai Tokkyo Koho JP 94-229,124
140. NuoFu C, Xingru Z, Lanying L, Xie X, Mian Z (2000) Mater Sci Eng B 75: 134
141. Moore AW (1990) J Cryst Growth 106: 6
142. Schwetz KA (1999) In: Ullmann's encyclopedia of industrial chemistry. Wiley-VCH, Weinheim, p 1
143. Cameron DC (1996) J Chem Vap Depos 3: 133
144. Okano T, Yamashita H (1998) Jpn Kokai Tokkyo Koho JP 10,134,815
145. Iwaki T, Niikura J, Gyoten H, Hosoi A (1986) Jpn Kokai Tokkyo Koho JP 61-248,363
146. Shaffer G, Hill RF (1999) Advanced Ceramics Corporation USA US 5,898,009
147. Nakao K (2000) Jpn Kokai Tokkyo Koho JP 2,000,086,443
148. Bartnitskaya TS, Ostrovskaya TM, Fenochka BV (1999) Powder Metall Met Ceram 38: 152
149. Bartnitskaya TS, Ostrovskaya NF, Vereshchaka VM, Kurdyumov AV (1999) Powder Metall Met Ceram 38: 240
150. Zhang G, Chen Z (1998) Gaofenzi Cailiao Kexue Yu Gongcheng 14: 94
151. Wakatsuki M, Ichinose K, Aoki T (1972) Mater Res Bull 7: 999
152. Vel L, Demazeau G, Etourneau J (1991) Mat Sci Eng B10: 149-164
153. Sumiya H, Iseki T, Onodera A (1983) Mater Res Bull 18: 1203
154. Gladskaya IS, Kremkova CN, Slesarev VN (1986) J Less-Common Met 117: 241
155. Sumiya H, Uesaka S, Satoh S (2000) J Materials Science 35: 1181-1186
156. Lux B, Kalss W, Haubner R, Taniguchi T (1999) Diamond Related Mater 8: 415-422
157. Wentorf RH Jr (1961) J Chem Phys 34: 809
158. DeVries RC, Fleischer JF (1972) J Cryst Growth 13/14: 88
159. Solozhenko VL, Turkevich VZ (1998) Diamond Related Mater 7: 43-46
160. Solozhenko VL, Turkevich VZ (1997) Mater Lett 32: 179-184
161. Endo T, Fukunaga O, Iwata M (1979) J Mater Sci 14: 1676
162. Lorenz H, Orgzall I (1995) Diamond Related Mater 4: 1046-1049
163. Lorenz H, Peun T, Orgzall I (1997) Appl Phys A 65: 487-497
164. Solozhenko VL, Turkevich VZ, Holzapfel WB (1999) J Phys Chem B 103: 8137-8140
165. Lorenz H, Orgzall I, Hinze E (1995) Diamond Related Mater 4: 1050-1055
166. Sumiya H, Tsuji K, Yazu S (1989) J Jpn Soc Powder Powder Metall 36: 752
167. Endo T, Fukunaga O, Iwata M (1981) J Mater Sci 16: 2227
168. Demazeau G, Biardeau G, Vel L (1990) CR Acad Sci II 310: 897
169. Demazeau G, Biardeau G, Vel L (1990) Mater Lett 10: 139
170. Gameza LM (2000) Poroshk Metall (Minsk) 22: 21-23
171. Susa K, Kobayashi T, Taniguchi S (1974) Mater Res Bull 9: 1443
172. Susa K, Kobayashi T, Taniguchi S (1976) High Temp High Press 8: 631
173. Kobayashi T, Susa K, Taniguchi S (1975) Mater Res Bull 10: 1231
174. Kobayashi T, Susa K, Taniguchi S (1977) Mater Res Bull 12: 847
175. Shulzhenko AA, Sokolov AN, Dub SN, Belyavina NN (2000) Sverkhtverd Mater 2: 30-35
176. Demazeau G, Gonnet V, Solozhenko V, Tanguy B, Montignaud H (1995) CR Acad Sci Ser II 320: 419-422
177. Solozhenko VL, Le Godec Y, Mezouar M, Besson J-M, Syfosse G (1999) ESRF Highlights 92-93
178. Taniguchi T, Yamaoka S (2001) J Cryst Growth 222: 549-557
179. Mishima O, Yamaoka S, Fukunaga O (1987) J Appl Phys 61: 2822
180. Taniguchi T, Yamaoka S (2000) New Diamond Front Carbon Technol 10: 291-299
181. Sawaoka A, Soma T, Saito S (1974) Jpn J Appl Phys 13: 891-892
182. Komatsu T, Nomura M, Kakudate Y, Fujiwara S (1998) J Chem Soc Faraday Trans 94: 1649-1655
183. Komatsu T (1998) J Chem Soc Faraday Trans 94: 101-104
184. Dischler B, Wild C (1998) Low pressure synthetic diamond, Springer, Berlin, Heidelberg, New York

185. Kalss W (1998) Doctoral Thesis, Vienna University of Technology
186. Sachdev H, Strauss M (2000) Diamond Related Mater 9: 614–619
187. Reinke S, Kuhr M, Kulisch W (1994) Diamond Related Mater 3: 341–345
188. Widany J, Weich F, Köhler T, Porezag D, Frauenheim T (1996) Diamond Related Mater 5: 1031–1041
189. Shtansky DV, Tsuda O, Ikuhara Y, Yoshida T (2000) Acta Mater 48: 3745–3759
190. McKenzie DR, McFall WD, Sainty WG, Davis CA, Collins RE (1993) Diamond Related Mater 2: 970–976
191. McKenzie DR (1993) J Vac Sci Technol B11: 1928–1935
192. Dworschak W, Jung K, Ehrhardt H (1995) Thin Solid Films 254: 65–74
193. Uhlmann S, Frauenheim T, Stephan U (1995) Phys Rev B51: 4541
194. Reinke S, Kuhr M, Kulisch W, Kassing R (1995) Diamond Related Mater 4: 272
195. Kester DJ, Messier R (1992) J Appl Phys 72: 504
196. Reinke S, Kuhr M, Kulisch W (1995) Surf Coat Technol 74/75: 723–728
197. Weissmantel C, Bewilogua K, Dietrich D, Erler HJ, Hinneberg HJ, Klose S, Nowick K, Reisse G (1980) Thin Solid Films 72: 19
198. Weissmantel C (1981) J Vac Sci Technol 18: 179–185
199. Kulisch W, Reinke S (1997) Diamond Films Technol 7: 105–138
200. Hofsäss H, Ronning C, Griesmeier U, Gross M, Reinke S, Kuhr M, Zweck J, Fischer R (1995) Nucl Instrum Methods Phys Res B 106: 153–158
201. Saitoh H, Yarbrough WA (1992) Diamond Related Mater 1: 137–146
202. Seidel KH, Reichelt K, Schaal W, Dimigen H (1987) Thin Solid Films 151: 243–249
203. Reisse G, Weissmantel S (1999) Thin Solid Films 355–356: 105–111
204. Satou M, Fujimoto F (1983) Jpn J Appl Phys 22: L171–L172
205. Setsuhara Y, Kumagai M, Suzuki M, Suzuki T, Miyake S (1999) Surf Coat Technol 116/119: 100–107
206. Boyen H-G, Widmayer P, Schwertberger D, Deyneka N, Ziemann P (2000) Appl Phys Lett 76: 709–711
207. Kulisch W, Freudenstein R, Klett A, Plass MF (2000) Thin Solid Films 377/378: 170–176
208. McKenzie DR, McFall WD, Smith H, Higgins B, Boswell RW, Durandet A, James BW, Falconer IS (1995) Nucl Instrum Methods Phys Res B 106: 90–95
209. Barth K-L, Neuffer A, Ulmer J, Lunk A (1996) Diamond Related Mater 5: 1270–1274
210. Barth K-L, Lunk A, Ulmer J (1997) SurfCoat Technol 92: 96–103
211. Goranchev B, Schmidt K, Reichelt K (1987) Thin Solid Films 149: L77–L80
212. Schütze A, Bewilouga K, Lüthje H, Kouptsidis S, Jäger S (1995) Surface Coatings Technol 74/75: 717–722
213. Lüthje H, Bewilouga K, Daaud S, Johansson M, Hultman L (1995) Thin Solid Films 257: 40–45
214. Deng J, Wang B, Tan L, Yan H, Chen G (2000) Thin Solid Films 368: 312–314
215. Kulikovsky V, Shaginyan LR, Vereschaka VM, Hatynenko NG (1995) Diamond Related Mater 4: 113–119
216. Ulrich S, Theel T, Schwan J, Ehrhardt H (1997) Surf Coat Technol 97: 45–59
217. Okamoto M, Yokoyama H, Osaka Y (1990) Jpn J Appl Phys 29: 930–933
218. Weber A, Bringmann U, Nikulski R, Klages CP (1993) Diamond Related Mater 2: 201
219. Chayahara A, Yokoyama H, Imura T, Osaka Y (1987) Jpn J Appl Phys 26: L1435–L1436
220. Yokoyama H, Okamoto M, Osaka Y (1991) Jpn J Appl Phys 30: 344–348
221. Ye M, Delplancke-Ogletree MP (2000) Diamond Related Mater 9: 1336–1341
222. Kuhr M, Reinke S, Kulisch W (1995) Surf Coat Technol 74/75: 806–812
223. Kuhr M, Reinke S, Kulisch W (1995) Diamond Related Mater 4: 375–380
224. Kouvetakis J, Patel VV, Miller CW, Beach DB (1990) J Vac Sci Technol A8: 3929–3933
225. Yuzuriha TH, Hess DW (1986) Thin Solid Films 140: 199–207
226. Karim MZ, Cameron DC, Hashmi MSJ (1991) Surf Coat Technol 49: 416–421
227. Karim MZ, Cameron DC, Hashmi MSJ (1994) Diamond Related Mater 3: 551–554
228. Saitoh H, Yarbrough WA (1991) Appl Phys Lett 58: 2228–2230
229. Saitoh H, Yarbrough WA (1991) Appl Phys Lett 58: 2482–2484

230. Matsumoto S, Zhang W (2001) New Diamond Front Carbon Technol 11: 1–10
231. Mendez JM, Muhl S, Farías M, Soto G, Cota-Araiza L (1991) Surf Coat Technol 41: 422–426
232. Mendez JM, Muhl S, Andrade E, Cota-Araiza L, Farías M, Soto G (1994) Diamond Related Mater 3: 831–835
233. Matsumoto S, Nishida N, Akashi K, Sugai K (1996) J Mater Sci 31: 713–720
234. Zhang WJ, Matsumoto S (2000) Chem Phys Lett 330: 243–248
235. Zhang WJ, Matsumoto S (2000) Appl Phys A 71: 469–472
236. Yap YK, Aoyama T, Wada Y, Yoshimura M, Mori Y, Sasaki T (2000) Diamond Related Mater 9: 592–595
237. Hollemann AF, Wiberg E (1995) Lehrbuch der anorganischen Chemie, 105th edn., Walter de Gruyter, Berlin
238. D'Evelyn MP, Taniguchi T (1999) Diamond Related Mater 8: 1522–1526
239. Sumiya H, Sato S, Yazu S (1994) US Pat US 5,332,629
240. Uesaka S, Sumiya H, Itozaki H, Shiraishi J, Tomita K, Nakai T (2000) SEI Tekunikaru Rebyu 156: 18–23
241. Sumiya H, Uesaka S (2000) New Diamond Front Carbon Technol 10: 40
242. Anonymous (1988) Int Steel Met Mag 26: 516–8
243. Novikov NV, Shulzhenko AA, Petrusha IA (1987) Sverkhtverd Mater 6: 3/8
244. Hara A, Yatsu S (1986) Jpn Kokai Tokkyo Koho JP 61,209,958
245. Hara A, Yatsu S (1987) Jpn Tokkyo Koho JP 77-32,600
246. Hooper RM, Brookes CA (1986) Inst Phys Conf Ser 75: 907–917
247. Hara A, Yatsu S (1987) Japan Kokai Tokkyo Koho 62-07,151
248. Kurokawa Y (1985) Jpn Kokai Tokkyo Koho JP 60,210,573
249. Cerceau JM, Hall HT (1986) Eur Pat Appl EP 181,258
250. Hasegawa H (1989) Jpn Kokai Tokkyo Koho JP 01,131,067
251. Nishigaki K, Kikuchi N, Miwa K (1980) Jpn Kokai Tokkyo Koho JP 55,015,963
252. Collier MW, Yao X, Bowers BG (2000) US Pat US 6,140,262
253. Hooper RM, Guillou MO, Henshall JL (1991) J Hard Mater 2: 223–31
254. Yazu S, Kohno Y, Sato S, Hara A (1981) Mod Dev Powder Metall 14: 363–71
255. Hasegawa H (1988) Jpn Kokai Tokkyo Koho JP 63,035,455
256. Benko E, Stanislaw JS, Krolicka B, Wyczesany A, Barr TL (1999) Diamond Related Mater 8: 1838–1846
257. Cerceau J-M, Boyat Y (1995) Fr Demande FR 2,713,223
258. Mitsusaka K, Yatsu S (1985) Jpn Kokai Tokkyo Koho JP 60,184,650
259. Kuratomi T (1985) Jpn Kokai Tokkyo Koho JP 60,230,956
260. Kuratomi T (1993) Jpn Kokai Tokkyo Koho JP 05,339,062
261. Benko E, Morgiel J, Czeppe T (1997) Ceram Int 23: 89–91
262. Morgiel J, Benko E (1995) Mater Lett 25: 49–52
263. Benko E, Morgiel J, Czeppe T, Barr T (1998) J Eur Ceram Soc 18: 389–393
264. Kuratomi T (1994) Jpn Kokai Tokkyo Koho JP 06,293,567
265. Kuratomi T (1992) Jpn Kokai Tokkyo Koho JP 04,144,966
266. Kuratomi T (1991) Jpn Kokai Tokkyo Koho JP 03,164,475
267. Kono Y, Hara A (1985) Jpn Kokai Tokkyo Koho JP 60,264,371
268. Nakai T, Hara A, Goto M (1987) Eur Pat Appl EP 223,585
269. Journal for grinding and cutting technologies (1997) Issue 123
270. Yokogawa M, Yokogawa K (1986) Werkstatt Betr 119: 788–794
271. Arnot RN, Fischbacher MJ (1995) VDI-Z 137(7/8): 36–38
272. Yokogawa M, Yokogawa K (1992) Int J Jpn Soc Precis Eng 26: 108–114
273. Nagata A (2001) Jpn Kokai Tokkyo Koho JP 2,001,062,732
274. Choe S, Sonn S (1997) Korea Patent KR 9,709,218
275. Viernekes N (1986) Sprechsaal 119: 1105–1106
276. Sato K, Nagata A (1990) Jpn Kokai Tokkyo Koho JP 02,055,272
277. Juricek V (1989) Czech Pat CS 257,551

278. Procyk B, Staniewicz-Brudnik B, Majewska-Albin K, Zawada A, Bieniarz P, Hohne D (2000) Interceram 49: 308–314
279. Ichikawa K, Ando D (1989) Jpn Kokai Tokkyo Koho JP 01,179,763
280. Maekawa K, Yokoyama Y, Ohshima I (2001) Key Eng Mater 196: 133–140
281. Omori O, Furukawa H (2001) Jpn Kokai Tokkyo Koho JP 2,001,062,721
282. Yoshioka H, Kodama Y, Shigemura S, Funada T, Morimoto S (1989) US Pat US 4,855,019
283. Hagiuda Y, Takaya M, Matsunaga M, Yasunaga N (1990) Hyomen Gijutsu 41: 852–853
284. Osaka Diamond (1983) Jpn Tokkyo Koho JP 58,002,034
285. Takeuchi K, Senba T (2000) Jpn Kokai Tokkyo Koho JP 2,000,254,866
286. Wentorf RH (1962) J Chem Phys 36: 1990–1991
287. Mishima O, Tanaka J, Yamaoka S (1987) Science 238: 181–183
288. Mishima O, Era K, Tanaka J, Yamaoka S (1988) Appl Phys Lett 53: 962–964
289. Hara A, Yatsu S (1987) Jpn Tokkyo Koho JP 62,007,151

Silicon Nitride Ceramics

G. Petzow[1], M. Herrmann[2]

[1] Max-Planck-Institut für Metallforschung, Heisenbergstr. 5, 70569 Stuttgart, Germany
 e-mail: petzow@aldix.mpi-stuttgart.mpg.de
[2] Fraunhofer Institut für Keramische Technologien und Sinterwerkstoffe,
 Winterbergstrasse 28, 01277 Dresden, Germany
 e-mail: mathias.herrmann@ikts.fhg.de

Silicon nitride ceramics is a generic term for a variety of alloys of Si_3N_4 with additional compounds necessary for a complete densification of the Si_3N_4 starting powder. They are heterogeneous, multicomponent materials characterised by the inherent properties of the crystalline modifications α and β of Si_3N_4 and the significant influence of the densification additives. With a view to ability of the α and β modification to form solid solutions α-Si_3N_4 (αss) and β-Si_3N_4 (βss) solid solutions can be distinguished. Each group contains engineered materials with interesting properties for special applications. Phase relations and microstructures determine the properties decisively. Composition of the phases, the distribution of the grains, their aspect ratio and the grain boundary phase are pronounced microstructural features. The formation of the microstructure strongly depends on the one hand on the quality of the Si_3N_4 starting powders, which closely is related to the chemistry of the production process, and on the other on the liquid phase sintering as the most important step in the densification route. The interrelation between pure Si_3N_4, the densification of the powder including the role of sintering additives, microstructural engineering, physicochemical properties of the sintered Si_3N_4 ceramics (SSN, GPSN, HPSN, HIP-SSN, HIP-SN) are described in more detail and compared to reaction bonded Si_3N_4 ceramics (RBSN), which are produced by nitridation of silicon powders.

Keywords: Silicon nitride ceramics, Phase relations, Processing, Microstructure, Properties

Structure and Bonding, Vol. 102
© Springer-Verlag Berlin Heidelberg 2002

Abbreviations

GPSN	gas-pressure sintered Si$_3$N$_4$ ceramics
HPSN	hot pressed Si$_3$N$_4$ ceramics
HIP-SN	encapsulated hot isostatically pressed Si$_3$N$_4$ ceramics
HIP-SSN	presintered hot isostatically pressed Si$_3$N$_4$ ceramics (also Sinter-HP-SN)
Sinter-HIP-SN	presintered hot isostatically pressed Si$_3$N$_4$ ceramics
SSN	sintered Si$_3$N$_4$ ceramics
RBSN	reaction bonded Si$_3$N$_4$ ceramics
SRBSN	sintered reaction bonded Si$_3$N$_4$ ceramics
HIP-SRBSN	hot isostatically pressed SRBSN ceramics
AAS	atomic absorbtion flame spectroscopy
AES	atomic emission spectroscopy
CAPLUS	data bank of Chemical Abstracts
CVD	chemical vapour deposition
E	Young's modulus
HRTM	high resolution transmission electron microscopy
HT	high temperature
HIP	hot isostatic pressing
K$_{IC}$	fracture toughness (critical stress intensity factor)
M	interstitial metal in αss (Eq. 6)
$M_i^{V^{i+}}$	cations with the charge V_i^+
Me	metals
MS	mass spectroscopy
OES	optical emission spectroscopy
RE	rare earth metals
RT	room temperature
SEM	scanning electron microscopy

SHS	self-propagating high temperature synthesis
T_g	transition temperature of the glassy phase
TEM	transmission electron microscopy
X_i^{Zi-}	anion with the charge Z_i^- (usually O^{2-} and N^{3-})
XRF	X-ray fluorescence analysis
XRD	X-ray diffraction
XPS	X-ray photoelectron spectroscopy
Z	atomic number
a	flaw size (Sect. 7.2)
a	thermal diffusivity (Sect. 7.1)
at% (M_i)	atomic % of cation $M_i^{V_i^+}$ with the charge V_i^+
at% (X_i)	atomic % of anion $X_i^{Z_i}$ with the charge Z_i^-
equ.%	equivalent %
m	$x \cdot V^+$
n	number of components (Sect. 3)
n	degree of nitrogen substitution in αss (Eq. 6; Sect. 3.3; 6.2; 7.1)
n	growth rate exponent (Sect. 7.2)
V^+	charge of cation M
x	amount of stabilising cation M
z	degree of substitution in βss ($Si_{6-z}Al_zN_{8-z}O_z$)
α	thermal expansion coefficient (Sect. 2; 7.1)
αss	solid solutions based on the α-Si_3N_4 modification
βss	solid solutions based on the β-Si_3N_4 modification
$\dot{\varepsilon}$	creep rate
ε	deformation during creep
ν	Poisson ratio
σ	bending or tensile strength
ζ	zeta-potential

1
Introduction

Silicon nitride has the composition Si_3N_4 and its chemical bonding is predominantly covalent. Si_3N_4 represents the backbone of silicon nitride ceramics, a class of ceramic materials which, because of their exceptional profile of properties, are gaining increasing acceptance in engineering applications.

Natural sources of Si_3N_4 are extremely rare and are a mineralogical curiosity that has no significance as a raw material. The mineral Nierite is named after the pioneer of mass spectroscopy, A.O.C. Nier, and is observed as an inclusion in meteorites. It consists chiefly of the α modification which is interspersed with inclusions of β-Si_3N_4 whiskers. The ratio of α to β in the different finds can show considerable variance [1, 2].

All silicon nitride ceramics are derived from synthetic materials, exclusively. The first report on the synthesis of Si_3N_4 was in 1859 by Sainte-Claire Deville and Wöhler [3]. Among the problems of greatest concern to chemists in those days was the utilisation of atmospheric nitrogen for agricultural and industrial purposes. In particular, there was a need for a highly effective fertiliser to improve agricultural yields. In this connection, the bonding of nitrogen to silicon became of great interest since it was discovered that silicon is able to take up nitrogen at high temperatures and to release it in the form of ammonia. This led to many investigations into the synthesis of nitrogen-silicon compounds [3–7]. It might be mentioned that all the synthesis methods used today in industry were developed in principle in the 19th century.

Unfortunately the dream of a new effective fertiliser from silicon nitride was not fulfilled. Because of its high chemical and thermal stability it does not decompose easily, which is an important precondition for fertilisers. The goal of synthesising nitrogen-silicon compounds finally lost its appeal when simpler and cheaper processes for the synthesis of ammonia were discovered [8]. On the other hand, the high stability – undesirable in those days – is a significant feature of silicon nitride as a basis for advanced ceramics which, because of their peculiar combination of properties, fill a gap in the spectrum of engineering materials. Soon after 1950 silicon nitride was being employed as refractories and nozzles in jet engines and rockets [9, 10] which are produced by the reaction bonding route, i.e., by direct nitridation of compacted silicon powder parts. At the same time, comprehensive studies were conducted on the dependence of the properties of these particular silicon nitride materials on process variables [11, 12].

The reaction bonded materials are not completely dense. For optimal utilisation of the inherent good properties of silicon nitride, the powders must be fully densified to compacts. This was not possible in those days because suitable densification techniques were lacking. Since Si_3N_4 under ambient conditions does not melt but decomposes into the elements at about 2120 K a densification by melting was not feasible. The production of high-density silicon nitride materials became possible only after the techniques of densification had matured. This happened in the early 1960s with the introduction of hot pressing, originally developed in powder metallurgy for densification of high-melting metallic powders under the simultaneous application of high pressure and high temperature. The hot pressing of silicon nitride powders led to further progress. Deeley, Herbert and Moore reported on sintering studies of silicon nitride powders with small amounts of various oxide and nitride additives and achieved bodies of almost theoretical density by hot pressing [13]. So it is not surprising that the first publication on fully densified silicon nitride qualities did not appear in a chemical or a ceramic periodical but in the journal "Powder Metallurgy".

Whilst today the densification of silicon nitride powders is possible by normal sintering without additional pressure, the additives in general are still a must. Only with high pressures and temperatures is densification without additives possible. As a consequence the resulting materials are not pure silicon nitride qualities but mixtures or alloys of silicon nitride with additional compounds necessary for a complete densification. These materials are generally named SILICON NITRIDE CERAMICS, which is a generic term used for a variety of types having different compositions. In other words: silicon nitride ceramics are multicomponent mixtures or alloys of Si_3N_4.

About one hundred years after Si_3N_4 was synthesised for the first time, dense compacts could finally be prepared from its powders. Thus did an obscure chemical compound become an engineering material. The significant stages of its development are mentioned in Fig. 1.

The spectrum of properties provided by dense Si_3N_4 ceramics has proved to be of exceptional value in numerous applications and improvements that were previously not considered feasible for lack of a suitable material. This has

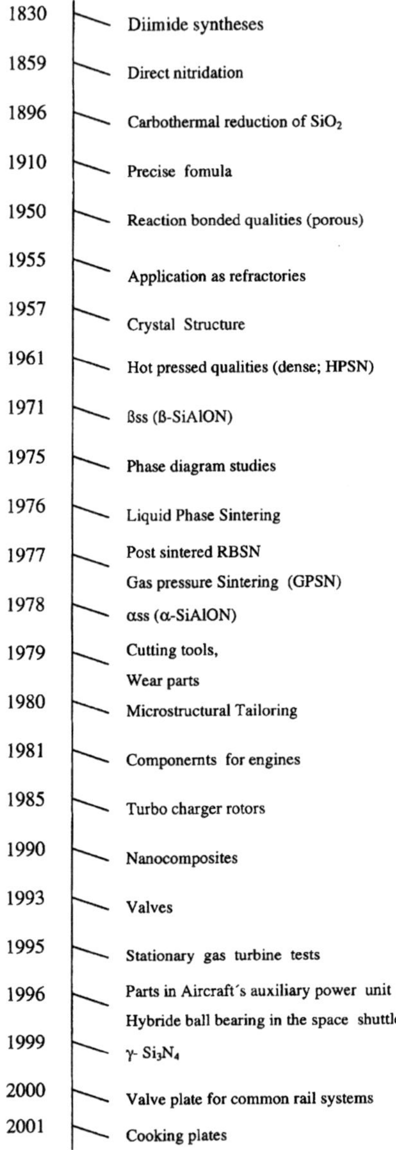

Fig. 1. Some historical data for the development of Si_3N_4 ceramics; a chemical compound becomes an advanced ceramic

generated a large body of systematic scientific and technological investigations that have led to a deeper understanding of this class of materials. The results and insights thereby gained are reflected in numerous reports and publications whose number, based on the CAPLUS data bank of Chemical Abstracts, is approaching fifty-five thousand. The continuing rise in the number of relevant publications on the subject of Si_3N_4 ceramics is illustrated in Fig. 2. But as yet, no comprehensive compilation of the information on this interesting and diverse class of materials is available in text book form, although several helpful review articles, summaries of detailed aspects and data collections can be found in the literature [14–21].

2
Crystalline Modifications

Three crystalline modifications of Si_3N_4 are known (Fig. 3a–c) [22–28]. The corresponding data are listed in Table 1 together with some characteristic properties. Whilst α- and β-Si_3N_4 are to be produced under normal nitrogen pressure, the recently observed γ modification is originated at high pressure and temperature [23, 29].

2.1
Crystal Structures and Inherent Properties

The **α modification** dominates in the commonly produced Si_3N_4 powders (Sect. 4).

The lattice parameters of the α phase depend on the oxygen content dissolved in the structure. With increasing amounts of oxygen the a-parameter

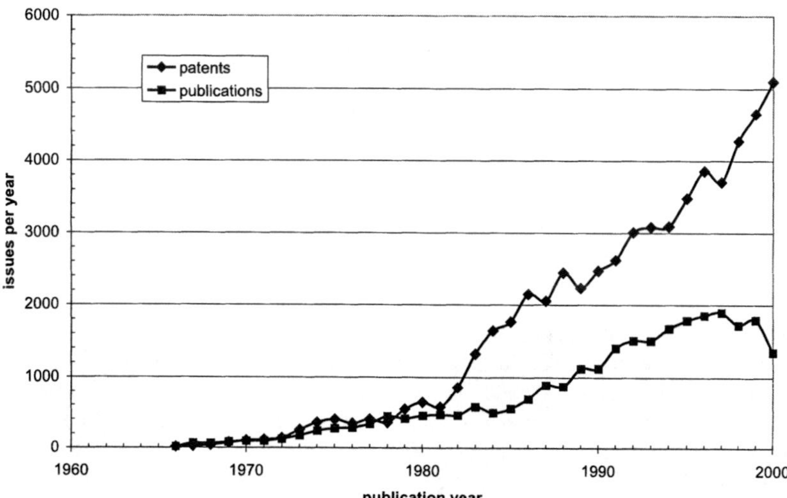

Fig. 2. The number of publications on Si_3N_4 ceramics since 1965 (source: Chemical Abstracts; CAPLUS)

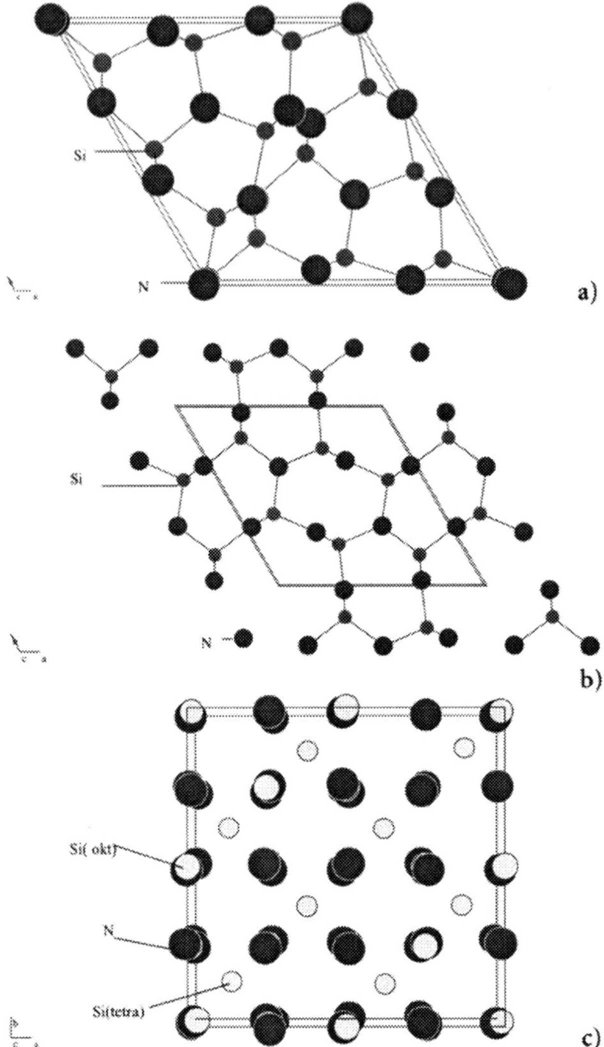

Fig. 3a, b. Projection of the crystal structures of α, β, γ (**a, b, c**)

decreases to 0.775 nm and the c-parameter increases to 0.5625 nm [22, 45]. From earlier investigations it is assumed that α is an oxide nitride with an oxygen content of 0.9–1.48 wt% [45]. (It should be taken into account that in ceramic literature "oxide nitride" is generally called "oxynitride".) However, more detailed investigations clearly demonstrate the existence of α-Si$_3$N$_4$ with oxygen solubility lower than 1 wt% in the absence of other impurities; in monocrystals up to 0.3 wt% have been measured [22]. The melting point of α-Si$_3$N$_4$ was detected at 2560 K under a nitrogen pressure of 120 MPa [46]. This value is more reasonable than that determined as 2200 K and 3.58 GPa [47].

Table 1. Crystal structures and properties of Si_3N_4 modifications

	α-Si_3N_4	β-Si_3N_4	γ-Si_3N_4
Space group	P31c; No. 159 [22]	$P6_3$; No. 173 [22, 25]	$Fd\bar{3}m$ [23, 26]
Lattice parameter			
a, nm	0.7818(3) [22]	0.7595(1) [25]	0.7738
c, nm	0.5591(4)	0.29023(6)	
Density, g cm^{-3} [22]	3.18	3.20	4,0 [26], 4.12 [29]
Coordinates of atoms	[24], [30]	[25], [45]	Table 2
Hardness	(110) plane: 2250 (HV 0.3)	(100) plane: 2100 (HV 0.025)	>30 GPa [23]
	(001) plane: 2200 (HV 0.3) [31]	(001) plane: 1326 (HV 0.025) [34]	
	(100) plane: 3660 (HV 0.3)		
	(001) plane: 2782 (HV 0.3) [32]	(100) plane: 2610 (HV 0.3)	
	(100) plane: 2830 (HV 0.3)	(001) plane: 1642 (HV 0.3) [32]	
	(001) plane: 2890 (HV 0.3) [33]		
Fracture toughness, MPa $m^{1/2}$	1.9–2.8 [31]		
Elastic constants	[calculated after 35]	[36]	
E_x, GPa	341	280	
E_z, GPa	343	540	
ν_{xy}	0.30	0.35	
ν_{zx}	0.29	0.25	
G_{xz}, GPa	132	124	
Standard molar enthalpies of formation	-828.9 ± 3.4 [37]	-827.8 ± 2.5 [37]	
$\Delta_f H_m^\circ$ KJ mol^{-1}	-850.9 ± 22.4 [38]	-852.0 ± 8.7 [38]	
Coefficient of thermal expansion 0–1000 °C, 10^{-6}/K	3.64 [39]	3.39 [39]	
α_a (0–500 °C)	2.72	2.01	
α_c-(0–500 °C)	3.14 [18a]	2.84 [18a]	
α_a (0–1000 °C)	3.61	3.23	
α_c-(0–1000 °C)	3.70	3.72	
Thermal conductivity, $W(mK)^{-1}$	110–150 [40]	[001] Direction 180	
		[100] Direction 68 [41]	
Si-diffusion, cm^2 s^{-1} [42]	1673 K: $0.45 \cdot 10^{-15}$		
	1773 K: $1 \cdot 10^{-15}$		
	Activation energy = 197 KJ mol^{-1}		
N-diffusion coefficient, cm^2 s^{-1} [43]	$D_{N\alpha} = 1.2 \cdot 10^{-12}$	$D_{N\beta} = 5.8 \cdot 10^6$	
	$Exp(-233$ KJ mol^{-1}/RT)	$Exp(-777$ KJ mol^{-1}/RT)	
Refractive index [44]	$n_o = 2.03$	$n_o = 2.02$	
	$n_e = 2.02$ (optically negative)	$n_e = 2.04$ (optically positive)	

These data seem unlikely because Si_3N_4 can be sintered at 1950–2000 °C (2225–2275 K) and 1–10 MPa nitrogen pressure without melting.

The **β-modification** is the main constituent of the majority of the Si_3N_4 ceramics. Different space groups were observed: the centrosymmetric $P6_3/m$ [45] and the corresponding non-centrosymmetric $P6_3$ [25, 35]. Detailed investigations on single crystals reveal $P6_3$. The atomic coordinates in the unit cell remain nearly constant up to 1633 K [48]. Deviations from the observed cell dimensions are presumably caused by aluminium and oxygen impurities (Sect. 3). The solubility of oxygen in the β structure is up to 0.258 wt% in the absence of other elements [49].

In perfect monocrystals of $β$-Si_3N_4 thermal conductivities up to 320 $W(mK)^{-1}$ are possible as shown by calculations [40]. This intrinsic thermal conductivity of Si_3N_4 is very near that of AlN [319 $W(mK)^{-1}$]. The deviations from these optimal values to the considerably lower values [<150 $W(mK)^{-1}$] obtained for real sintered Si_3N_4 ceramics are caused by the grain boundary phases and their distribution [40]. It is supposed that the thermal conductivity of Si_3N_4 can be improved by reduction of the oxygen dissolved in the lattice, as known for AlN ceramics. Compared to AlN the lower diffusion coefficient of oxygen in $β$-Si_3N_4 causes a much longer heat treatment to attain oxygen-free materials [40].

The **cubic γ-modification** has been recently observed under a pressure of 15 GPa and temperatures above 2000 K by the laser heating technique in a diamond cell [23] and in shock-wave compression experiments with pressures >33 GPa at 1800 K and >50 GPa at 2400 K [29]. This modification is often designated as the c-modification in the literature in analogy to the cubic boron nitride (c-BN). It has a spinel-type structure in which two silicon atoms are octahedrally coordinated by six nitrogen atoms, one silicon atom is coordinated tetrahedrally by four nitrogen atoms (Fig. 3c). The atomic coordinates for the cubic modification are given in Table 2. From calculations it is shown that this structure should have a high hardness similar to that of diamond and c-BN [23].

2.2
$α/β$ Relations

Both modifications are based on SiN_4 tetrahedra connected at the corners. The dimension of the SiN_4 tetrahedron is very similar to that of SiO_4. Therefore

Table 2. Atomic co-ordinates of $γ$-Si_3N_4 with space group $Fd\bar{3}m$; a = 0.77381(2) nm [Int. Tables No. 227) [26]

Position	x/a	y/a	z/a
Si_{tet} in 8(a)	1/8	1/8	1/8
Si_{oct} in 16(d)	1/2	1/2	1/2
N in 32(e)	0.25968(1)	0.25968(1)	0.25968(1)

some oxide nitrides exist containing both, SiO_4 and SiN_4 tetrahedra [50]. In the α- as well as in the β-modification each nitrogen belongs to 3 tetrahedra, i.e., the three-dimensional arrangement is a network of tetrahedra. Whilst in β only one layer of SiN_4 tetrahedra exists, α-Si_3N_4 has two layers shifted with respect to each other. This leads to a doubling of the c-parameter in the α-Si_3N_4 unit cell (Figs. 3a and b). The defects in the α- as well as in the β-structure are similar; in both cases the most commonly observed dislocations have a <0001> type Burgers vector [22, 51, 52].

The α-phase is metastable during sintering (e.g. at 1673–2273 K and 0.1 to 100 MPa N_2 pressure) and transforms irreversibly to β-Si_3N_4. Extensive studies have been made to determine the kinetics and mechanisms of the transformation [22, 39, 53]. A heat treatment of sintering additive-free α-Si_3N_4 powder at 1900°C results in a slow phase transformation and an epitaxial orientation of the α- and β-modifications in the c-direction [53a]. A wide range of data about the difference of the standard molar enthalpy of formation $\Delta_f H_m^\circ$ of the two modifications are found in the literature [25, 54], earlier data indicating that α-Si_3N_4 has a lower standard molar enthalpy of formation by several KJ in comparison to β [25] and therefore α-Si_3N_4 is the stable modification at room temperature [54]. These quite large deviations are very likely caused by the low qualities of the available Si_3N_4 powders containing different amounts of oxygen, carbon and other impurities, and in all cases the investigations were done on α/β mixtures. In recent investigations on high purity powders also different assumptions have been made on the chemical bonding of the impurities, e.g., carbon impurities were considered as SiC, oxygen as SiO_2 [37, 38]. These investigations have shown $\Delta_f H_m^\circ$ of the pure α and β to differ by only 1 KJ mol^{-1}, which is lower than the error of determination of $\Delta_f H_m^\circ$ (Table 1).

The most recent measurements of $\Delta_f H_m^\circ$ show that dissolved oxygen in the α-structure leads to a slight destabilisation [38]. The observed destabilisation contradicted the previous work reporting that the α-modification is stabilised by dissolved oxygen [45]. The recently observed values for $\Delta_f H_m^\circ$ are more negative than those given in compilations of chemical thermodynamic functions for inorganic substances, e.g., the JANAF tables which recommend $\Delta_f H_m^\circ = -744.75 \pm 29.3$ KJ mol^{-1} [55].

The β-Si_3N_4 structure exhibits channels parallel to the c-axis which are about 0.5 nm in diameter causing the relative high diffusion coefficient compared to α which do not show such channels (Table 1). These channels are changed into voids with seven nearest nitrogen neighbours in the α structure.

The hardness values of the crystals are given in Table 1. These data show that β has a much lower hardness than α. Also a high anisotropy of hardness and elastic properties are found for β, which is less pronounced in α.

Ab-initio calculations of the α- and β-Si_3N_4 structures show that the elastic constants of the α-structure are less anisotropic than those of the β-structure [35]. For both modifications the infrared absorption Raman spectra are reviewed [18b, 35]. Both α and β exhibit a more or less pronounced solid solubility for other elements. This phenomenon is treated in Sect. 3.

Table 3. Comprehensive overview of investigated Si_3N_4 ceramic systems and quasi-systems

Systems and quasi-systems	References
2-component systems	
Si-N	[60–62]
	[Structure and Bonding, Vol. 101 (2002) p 1–58]
Si-O	[61, 63–67]
3-component system and quasi-binary	
Si-N-O	[7, 60, 61, 66–69]
Si_3N_4-3(SiO_2)	
4-component systems and quasi-ternaries	
Si-N-O-Al	[70–88]
Si_3N_4-4(AlN)-2(Al_2O_3)-3(SiO_2)	
Si-N-O-Be	[70, 82, 83, 89–91]
Si_3N_4-4(Be_3N_2)-6(BeO)-3(SiO_2)	
Si-N-O-C	[60, 61, 67, 92–96]
Si-N-O-Ca	[82]
Si_3N_4-2(Ca_3N_2)-6(CaO)-3(SiO_2)	
Si-N-O-Ce	[97, 98]
Si_3N_4-4(CeN)-2(Ce_2O_3)-3(SiO_2)	
Si-N-O-Dy	[99]
Si_3N_4-4(DyN)-2(Dy_2O_3)-3(SiO_2)	
Si-N-O-Er	[99]
Si_3N_4-4(ErN)-2(Er_2O_3)-3(SiO_2)	
Si-N-O-Gd	[99–101]
Si_3N_4-4(GdN)-2(Gd_2O_3)-3(SiO_2)	
Si-N-O-La	[99, 101, 102]
Si_3N_4-4(LaN)-2(La_2O_3)-3(SiO_2)	
Si-N-O-Mg	[68, 69, 82, 84, 91, 98, 103–105]
Si_3N_4-2(Mg_3N_2)-6(MgO)-3(SiO_2)	
Si-N-O-Nd	[99–101, 106, 107]
Si_3N_4-4(NdN)-2(Nd_2O_3)-3(SiO_2)	
Si-N-O-Sm	[99]
Si_3N_4-4(SmN)-2(Sm_2O_3)-3(SiO_2)	
Si-N-O-Th	[82, 108]
Si_3N_4-Th_3N_4-3(ThO_2)-3(SiO_2)	
Si-N-O-Y	[82, 99, 109–112]
Si_3N_4-4(YN)-2(Y_2O_3)-3(SiO_2)	
Si-N-O-Yb	[99, 113–116]
Si_3N_4-4(YbN)-2(Yb_2O_3)-3(SiO_2)	
Si-N-O-Zr	[82, 109, 114]
5-component systems and quasi-quaternaries	
Si-N-O-Al-Be	[70, 82]
Si_3N_4-4(AlN)-2(Al_2O_3)-3(SiO_2)-6(BeO)-2(Be_3N_2)	
Si-N-O-Al-Ca	[82, 117–128]
Si_3N_4-4(AlN)-2(Al_2O_3)-3(SiO_2)-6(CaO)-2(Ca_3N_2)	
Si-N-O-Al-Ce	[129–131]
Si_3N_4-4(AlN)-2(Al_2O_3)-3(SiO_2)-2(Ce_2O_3)-4(CeN)	
Si-N-O-Al-Dy	[100, 132–138]
Si_3N_4-4(AlN)-2(Al_2O_3)-3(SiO_2)-2(Dy_2O_3)-4(DyN)	
Si-N-O-Al-Er	[100, 136]
Si_3N_4-4(AlN)-2(Al_2O_3)-3(SiO_2)-2(Er_2O_3)-4(ErN)	

Systems and quasi-systems	References
Si-N-O-Al-Gd Si_3N_4-4(AlN)-2(Al_2O_3)-3(SiO_2)-2(Gd_2O_3)-4(GdN)	[100, 133, 136, 139]
Si-N-O-Al-La Si_3N_4-4(AlN)-2(Al_2O_3)-3(SiO_2)-2(La_2O_3)-4(LaN)	[130, 131, 140]
Si-N-O-Al-Li	[71, 91, 141]
Si-N-O-Al-Mg Si_3N_4-4(AlN)-2(Al_2O_3)-3(SiO_2)-6(MgO)-2(Mg_3N_2)	[71, 82, 83, 98, 109, 142, 143]
Si-N-O-Al-Nd Si_3N_4-4(AlN)-2(Al_2O_3)-3(SiO_2)-2(Nd_2O_3)-4(NdN)	[88, 99–101, 107, 126, 132, 133, 135, 136, 140, 144–152]
Si-N-O-Al-Sm Si_3N_4-4(AlN)-2(Al_2O_3)-3(SiO_2)-2(Sm_2O_3)-4(SmN)	[100, 126, 132–134, 136, 138, 145, 150, 153, 154]
Si-N-O-Al-Y Si_3N_4-4(AlN)-2(Al_2O_3)-3(SiO_2)-2(Y_2O_3)-4(YN)	[82, 83,88, 98, 106, 110, 111, 127, 132, 133, 135, 147, 148, 150, 152, 154–170]
Si-N-O-Al-Yb Si_3N_4-4(AlN)-2(Al_2O_3)-3(SiO_2)-2(Yb_2O_3)-4(YbN)	[100, 116, 135, 136, 138, 147, 150, 171]
Si-N-O-Al-Zr	[71, 82, 172]
Si-N-O-C-Ti	[173]
Si-N-O-Ca-Mg Si_3N_4-2(Ca_3N_2)-2(Mg_3N_2)-3(SiO_2)-6(CaO)-6(MgO)	[174]
Si-N-O-La-Y Si_3N_4-4(LaN)-4(YN)-2(La_2O_3)-2(Y_2O_3)-3(SiO_2)	[175]
Si-N-O-Mg-Y Si_3N_4-2(Mg_3N_2)-4(YN)-2(Y_2O_3)-6(MgO)-3(SiO_2)	[176]
6-component systems	
Si-N-O-Al-B-C	[177]
Si-N-O-Al-Ca-La	[178]
Si-N-O-Al-Ca-Mg	[178]
Si-N-O-Al-Sm-Y	[154]
Si-N-O-Ar-C-H	[179]
7-component systems	
Si-N-O-Al-C-Mg-Y	[180]

3
Phase Diagrams

Phase diagrams are concise plots of equilibrium relationships for understanding heterogeneous materials. By their nature they only represent thermodynamic equilibrium conditions, relating the physical state of a mixture to the number of substances of which it is composed and the environmental conditions imposed on it. In other words: phase diagrams are comprehensive descriptions of the constitution of matter as far as relations between different phases are concerned. The principles of phase equilibria are central to an understanding of many scientific and technological disciplines and provide important guidelines in the production, processing, and applications of materials in general and especially of Si_3N_4 ceramics, which are related to heterogeneous systems of higher order, i.e., containing more than two components n.

In Table 3 all the systems are compiled on which the known Si_3N_4 ceramics are based, arranged according to the number of elements (components) involved. From the binary and ternary subsystems (n ≤ 3) only Si-N, Si-O and Si-N-O are listed, because they are the only ones of all the subsystems which are existent in the higher order systems of all Si_3N_4 ceramics. All other side systems (subsystems), whose components may affect the phase diagrams and properties of Si_3N_4 ceramics to varying degrees are not depicted in the interest of space. The pertinent metal oxide (Si-Me-O), metal nitride (Si-Me-N), and metal oxide nitride (Me-N-O) systems are generally more or less well-known and described elsewhere [56–59].

3.1
Thermodynamics and Phase Diagrams

In general, truly thorough investigations of systems are limited to a small number of areas of special scientific or technological interest. Often, there are only a few experimental results and thermodynamic estimates, and in some cases, the treatment and assumptions are purely speculative. This applies in particular for the vapour phase which is of great importance with respect to the operating conditions on the materials, such as sintering behaviour, vaporisation, oxidation, corrosion, etc. Almost all the thermodynamically calculated phase diagrams include the vapour phase which gives significant additional information to that gained by experiments, usually carried out under reducing conditions in excess nitrogen atmosphere. But in some cases the vapour phase is not or only partially considered which to some extent explains the different data and inconsistencies in the literature. In Table 3 the literature is compiled which includes descriptions derived mainly from controlled nitrogen (N_2) conditions or is a calculated result at a fixed nitrogen potential.

3.1.1
Si-N-O System

Silicon and Si_3N_4 have a high affinity for oxygen, which it contains both in solution and in adsorbed form. Analytically it is possible to distinguish between the two oxygen species, bulk or surface oxygen, respectively (Sect. 4.2). To understand the role of nitrogen and oxygen in the phase relations of Si_3N_4 ceramics, the binary systems Si-N and Si-O as well as the ternary system Si-N-O are of essential value. There exist both experimental data and solid thermodynamic calculations, which are based on critical evaluations of several of those experimental investigations and predictions using thermodynamic modelling. For the side system N-O the gas phase only is considered. The phase relations in the system Si-N-O depend sensitively on the partial pressures of N_2, O_2, SiO and temperature as shown in Fig. 4 for example. Therefore, the knowledge of the partial pressures of the vapour species N_2, O_2 and SiO is a valuable tool.

In the system Si-N-O a quasi-binary relation exists between the two stable compounds Si_3N_4 and SiO_2 of the two binary systems Si-N and Si-O.

In respect to Si_3N_4 ceramics the quasi-binary system Si_3N_4-SiO_2 is of greatest importance (Fig. 5). However, if the vapour phase is included in the

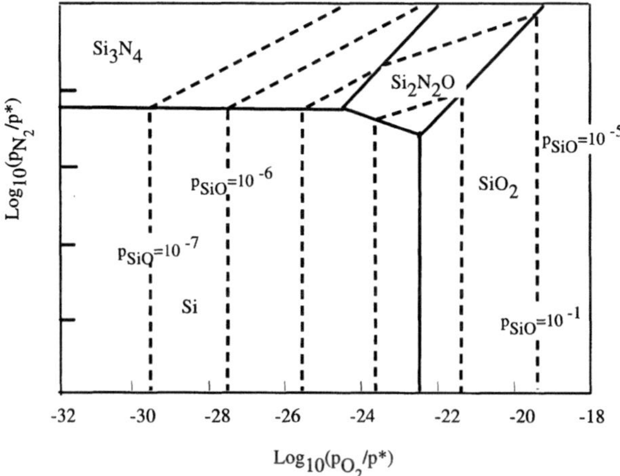

Fig. 4. Potential Si-N-O phase diagram at 1500 K including SiO isobares [66]; the pressures are given in bar

phase equilibria, this system can no longer be treated as a quasi-binary system since the composition of the vapour phase and the metallic melt is not restricted to compositions within the quasi-binary as is taken into account above 2100 K in Fig. 5. The decomposition of the ternary phase Si$_2$N$_2$O has been calculated to occur at 2137 K, according to the equation:

$$Si_2N_2O \Leftrightarrow N_2 + SiO + Si(l) \tag{1}$$

As can be concluded from Eq. (1), the decomposition will be significantly reduced with increasing N$_2$ pressure. Therefore at high N$_2$ pressure, the condensed phases can be treated as a quasi-binary system up to higher temperatures.

3.1.2
Si-N-O-Additives Systems

As mentioned before, the systems on which (technologically useful) Si$_3$N$_4$ ceramics are based, are without exception multicomponent, having n > 3 because in the production process (Sect. 5) sintering additives are necessary, which remain in the final product. The phase relations of such systems and their dependence on temperature and pressure can be represented in polyhedrons of three- or multi-dimensional order by introducing restricting conditions. Therefore phase diagrams of four-component systems become extremely cumbersome and those of more than four components are complex to the point of becoming intractable.

Fortunately the graphical representation of many phase relations in the Si$_3$N$_4$ ceramics can be simplified by conditions which lower the degree of freedom. Such a condition exists for reciprocal systems in which a double exchange reaction occurs

$$AX + BY \Leftrightarrow AY + BX \tag{2}$$

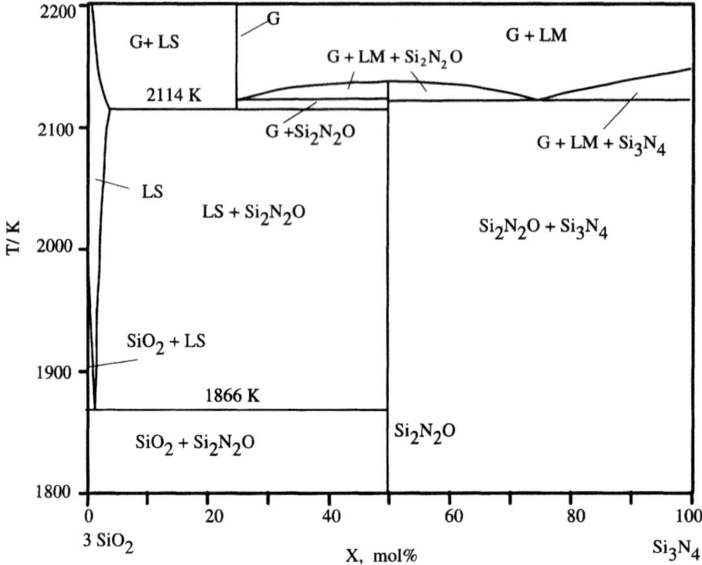

Fig. 5. Temperature-concentration section (isopleth) through the Si-N-O phase diagram from SiO$_2$ to Si$_3$N$_4$ [69]. Below 2114 K it is a quasi-binary system. G = gas phase; LS = oxide nitride liquid; LM = metallic liquid

where A and B are cations and X and Y anions. This reaction reduces the number of independent components to three and, thus, a two-dimensional representation of the concentrations is feasible. The requirement of electro-neutrality and the constant valence state of the components, labelled in equivalent percent (equ.%), result in a representation of the system on a square plane (isobaric and isothermal) with the four compounds at the corners, and the system can be treated as a quasi-ternary system with the same topological elements as all ternary systems. Five-component systems (quinary systems) can be treated in an analogous fashion if conditions between the involved compounds are quaternary and valency states of the elements are fixed. The resulting reciprocal salt system is a triangular prism, the Jänecke prism, named after E. Jänecke, a pioneer in the application of phase diagrams [181]. The conversion of at% into the less familiar equ.% is given by:

$$\text{equ.\%M}_i^{V_i^+} = 100\left(\frac{V_i \cdot \text{at\%}(M_i^{V_i^+})}{\sum_{i=1}^{n} V_i \text{at\%}(M_i^{V_i^+})}\right)$$

$$\text{and} \quad \text{equ.\%X}_i^{Z_i^-} = 100\left(\frac{Z_i \text{at\%}(X_i^{Z_i^-})}{\sum_{i=1}^{m} Z_i \text{at\%}(X_i^{Z_i^-})}\right) \tag{3}$$

$M_i^{V_i^+}$ are cations with the charge V_i^+; and at% (M_i) are the atomic % of the cation $M_i^{V_i^+}$. $X_i^{Z_i^-}$ are anions the charge z_i^- (usually O^{2-} and N^{3-}); and at% (X_i) is the atomic % of the anion $X_i^{Z_i-}$.

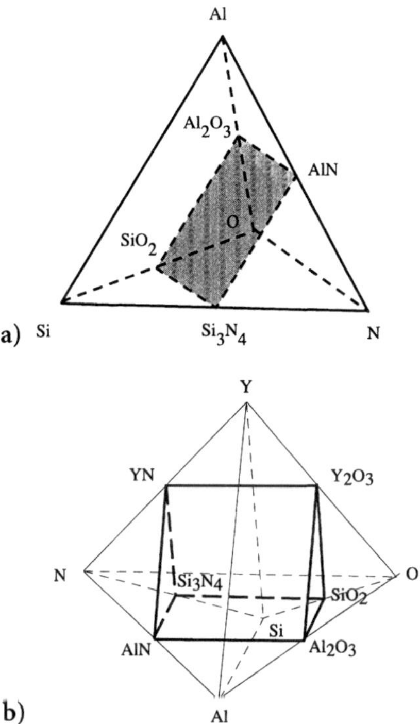

Fig. 6a, b. Derivation of quasi-subsystems in higher order systems (schematic). a) The quasi-ternary subsystem Si_3N_4-AlN-Al_2O_3-SiO_2 in the 4-component system Si-N-O-Al. b) The quasi-quaternary subsystem Si_3N_4-AlN-Al_2O_3-SiO_2-Y_2O_3-YN (Jänecke prism) of the 5-component system Si-N-O-Al-Y

In Fig. 6a the derivation of the square diagram Si_3N_4-AlN-Al_2O_3-SiO_2 in the quaternary system Si-N-O-Al is schematically explained and in Fig. 6b the Jänecke prism Si_3N_4-AlN-Al_2O_3-SiO_2-Y_2O_3-YN in the quinary system Si-N-O-Al-Y is derived.

The Si_3N_4 ceramics which are arranged in Table 3 according to the number of the involved components show remarkably different characteristics in their solution behaviour for additional elements. With a view to their ability to form solid solutions two groups can be distinguished:

1) β-Si_3N_4 solid solutions (βss) and
2) α-Si_3N_4 solid solutions (αss).

Each group contains engineered materials for special applications. Out of each group examples will be discussed in the following.

3.2
β-Si_3N_4 Solid Solutions (βss)

The solid solution capacity of the β-phase differs drastically depending on the dissolved elements; remarkable differences in the processing behaviour and properties are the consequence. Therefore it seems useful to subdivide the solubility of β into extended and low solid solutions.

3.2.1
Extended β-Si$_3$N$_4$ Solid Solutions

Aluminium and beryllium can be dissolved extensively in β-Si$_3$N$_4$. In these cases the Si^4 and N^{3-} ions may be exchanged by the other metals and oxygen, respectively. The exchange appears as a coupled substitution for reason of the valence electron balance occasioned by the stoichiometry. As can be concluded by the high number of citations (Table 3) the system Si-N-O-Al, usually named SiAlON, has found considerable interest and has been more intensively studied than other Si$_3$N$_4$ ceramics. It has a certain technological and scientific significance and therefore is chosen as an example for the extended β-solid solution group. In the many publications on these materials, several abbreviations are used to designate the different phases. Here we follow the recommendations given by IUPAC [182]. Incidentally, the expression SiAlON is often used as a generic name for Si$_3$N$_4$ ceramics, even when they do not contain Al. To avoid confusion, the relevant expressions α- and β-solid solutions are to be preferred.

All compositions in the system can be represented by a regular tetrahedron (Fig. 6a). Since the elements have their normal valencies (Si^{4+}, Al^{3+}, O^{2-}, N^{3-}) and the compositions are in equivalents all possible solid compounds lie on a square plane (shaded in Fig. 6a) in which any point represents a combination of 12 positive and 12 negative valencies. Therefore the square plane with Si$_3$N$_4$, AlN, Al$_2$O$_3$, and SiO$_2$ at its corners can be treated as a reciprocal salt system with the double exchange reaction:

$$Si_3N_4 + 2Al_2O_3 \Leftrightarrow 4AlN + 3SiO_2 \tag{4}$$

The corresponding phase relations can be represented in a quasi-ternary system with the four compounds as components [181].

Many versions of phase diagrams of the subsystem Si$_3$N$_4$-4(AlN)-2(Al$_2$O$_3$)-3 (SiO$_2$) have been published to date, based on experimental studies (Table 3). In the early periods the diagrams were reported as ternary systems Si$_3$N$_4$-AlN-Al$_2$O$_3$ [183] and Si$_3$N$_4$-SiO$_2$-Al$_2$O$_3$ [184]. But as mentioned previously, the treatment as a quasi-ternary reciprocal salt system is by far more appropriate. Based on more recent experimental and thermodynamic data in the side systems Al$_2$O$_3$-AlN [185] and improved modelling [186], it became possible to perform thermodynamic calculations which have secured the knowledge of the quasi-ternary system. In Fig. 7 one example is selected from many others to illustrate a calculated isothermal section at 1873 K and compare it with the experimental data. Other important intermediate phases in the system are Si$_2$N$_2$O (O), Si$_{12}$Al$_{18}$O$_{39}$N$_8$ (X) and six phases near the AlN corner [83]. These are interpreted in terms of AlN polytypoids and are not indicated in Fig. 7.

In β-Si$_3$N$_4$ the Si and N ions can be replaced by Al and O ions to form an extended solid solution according to the formula

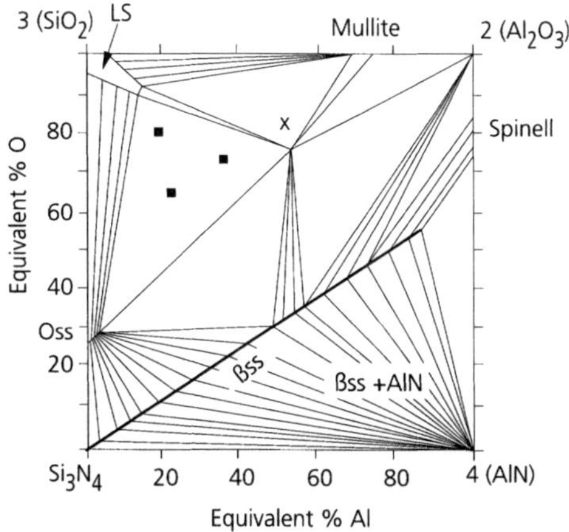

Fig. 7. Calculated isothermal section at 1873 K of the subsystem Si$_3$N$_4$-4(AlN)-2(Al$_2$O$_3$)-3(SiO$_2$) [77], compared with experimental data (■). LS = oxide nitride liquid; O = Si$_2$N$_2$O; X = Si$_{12}$Al$_{18}$O$_{39}$N$_8$; Mullite = Al$_2$Si$_3$O$_{12}$; Spinel = Al$_3$O$_3$N

$$Si_{6-z}Al_zO_zN_{8-z} \tag{5}$$

where z represents the number of replaced Al and O ions, respectively. The values of z reaches a maximum between 4 and 5 at about 2000 K but decreases with decreasing temperature. Full solubility occurs only with a constant cation: anion ratio of 3:4. This phase is strictly semistoichiometric, and compositional variations within this ratio do not require vacancies or interstitial atoms in the β-Si$_3$N$_4$ structure. In Table 4 the dependence of the α and β lattice parameters of the substitution of Si and N ions by Al and O ions is given.

Under high pressures cubic γss (Sect. 2.1) with z values (Eq. 5) up to 2.8 have been observed [186a].

3.2.2
Low β-Si$_3$N$_4$ Solid Solutions

The mentioned SiAlONs (and SiBeONs) are notable exceptions to the extended solubility of the β phase, whereas the solubility of all other alloying elements listed in Table 3 is quite limited. The related systems are important for an understanding of the effect of sintering additives.

Yttria is an often used additive to improve the sintering behaviour of Si$_3$N$_4$. In the quaternary system Si-N-O-Y the quasi-ternary system Si$_3$N$_4$-4(YN)-2(Y$_2$O$_3$)-3(SiO$_2$) exists. In Fig. 8a and b an isothermal section at 2000 K is shown as an example. Four oxide nitrides are existent: Y$_2$Si$_3$O$_3$N$_4$, YSiO$_2$N, Y$_4$Si$_2$O$_7$N$_2$, and Y$_{10}$(SiO$_4$)$_6$N$_2$, which are isotypic with the silicates mellilite, wollastonite, woehlerite and the phosphate apatite, respectively. Fig. 8a is the latest version of the diagram and differs slightly from earlier ones. For the purpose of sintering, only compositions in the compatibility

Table 4. Dependence of lattice parameters in α- and β-solid solutions on composition

Phase	a, nm	c, nm	Lit.
βss	$0.7601 + 0.00304z$	$0.2906 + 0.002554z$	87
$Si_{6-z}Al_zN_{8-z}O_z$	$0.7603 + 0.002967z$	$0.2907 + 0.002554z$	88
	$0.76069 + 0.00279z$	$0.29068 + 0.00263z$	84
αss			
$MSi_{12-n-m}Al_{m+N}N_{16-n}O_n$	$0.7752 + 0.0036m + 0.002n$	$0.5620 + 0.0031m + 0.004n$	127
M = Y	$0.7752 + 0.0045m + 0.0009n$	$0.5620 + 0.0048m + 0.0009n$	155
αss			
$MSi_{12-n-m}Al_{m+n}N_{16-n}O_n$	$0.7749 + 0.00673m + 0.00023n$	$0.5632 + 0.0055m + 0.00054n$	118
M = Ca			

triangles on either side of the Si_3N_4-$Y_2Si_2O_7$ join are essential. In this region there are no differences among the diagrams published.

For compositions on the Si_3N_4-$Y_2Si_2O_7$ tie line, a liquid forms at the binary eutectic temperature below 1550 °C and crystallises at 1500 °C. Compositions with higher Y_2O_3 contents than found on the join will fall into the compatibility triangle Si_3N_4-$Y_2Si_2O_7$-$Y_{10}(SiO_4)_6N_2$, which has a lower eutectic temperature than the triangle outlined by Si_3N_4-Si_2N_2O-$Y_2Si_2O_7$. Compositions with lower Y_2O_3 contents fall into the compatibility triangle outlined by Si_3N_4-$Y_2Si_2O_7$-Si_2N_2O.

The marked area of the Si_3N_4 corner in Fig. 8a indicates the composition range of the useful Si_3N_4 ceramics and is enlarged in Fig. 8b. The narrow arrangement of phase regions that fans out of the Si_3N_4 corner shows clearly that small variations in composition, as occur readily in the fabrication and processing of Si_3N_4 ceramics, may have very serious consequences for the composition of phases and the ensuing properties. In this connection it becomes obvious that the type of crucibles, powder beds and furnace atmospheres chosen for sintering are crucial for the physico-chemical and mechanical properties of the materials produced. These problems are exacerbated by the kinetic retardation of the equilibria, which is especially true of the intermediate oxide nitride $Y_{10}(SiO_4)_6N_2$.

The systems with low βss have been studied far less than Si-Al-O-N, except those without Al_2O_3 as a second additive. However, most of the commercial Si_3N_4 ceramics do contain Al_2O_3 as sintering aid, usually as a mixture with others. But the resulting βss have only small values of z. It is considered that Si_3N_4 ceramics with z-values ≥ 0.5 belong to the group of extended βss and all others with z-values <0.5 belong to the group of low βss [15]. This separation appears arbitrary, but is justified by the marked differences in the composition, corresponding properties, and by differences in the processing. To produce Si_3N_4 ceramics with $z \geq 0.5$ usually AlN is added to the presintered powder mixture. Therefore this distinction will be retained herein.

Usually the sintering aids are low melting mixtures of two or more oxides. As a consequence the related systems become more complex. As can be seen from Table 3, often mixtures of Al_2O_3 and another oxide are common, as for instance Al_2O_3 with Y_2O_3. They form together the five-

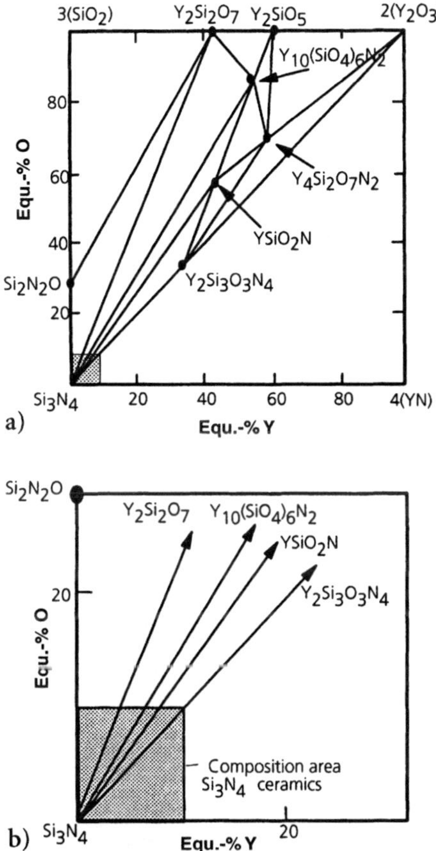

Fig. 8a, b. The subsystem Si_3N_4-4(YN)-2(Y_2O_3)-3(SiO_2) [111]. a) Isothermal section at 1773 K. b) Si_3N_4-rich area of a

component system Si-N-O-Al-Y, which can be reduced to a quasi-quaternary system according to the schematic Fig. 6b, the resulting Jänecke prism. It consists of three quasi-ternary reciprocal salt systems on the square plans Si_3N_4-4(AlN)-2(Al_2O_3)-3(SiO_2), Si_3N_4-4(YN)-2(Y_2O_3)-3(SiO_2), 4(AlN)-2(Al_2O_3)-2(Y_2O_3)-4(YN) and two ternary systems on the triangular faces Si_3N_4-4(AlN)-4(YN), 2(Al_2O_3)-2(Y_2O_3)-3(SiO_2). These relations make the complex diagrams of the Si_3N_4 ceramics appear simpler and easier to depict, but without necessarily making them more accessible. Detailed information from the three-dimensional figure can be obtained from two-dimensional plots of compatibility triangles. This is demonstrated with the system Si_3N_4-4(AlN)-2(Al_2O_3)-3(SiO_2)-2(Y_2O_3)-4(YN) in Fig. 9. In this Jänecke prism for reasons of clarity only phase relations of the Si_3N_4-4(AlN)-2(Al_2O_3)-3(SiO_2) plane and the subsystem Si_3N_4-YN·3AlN-$^4/_3$(AlN·Al_2O_3) (shaded) are represented.

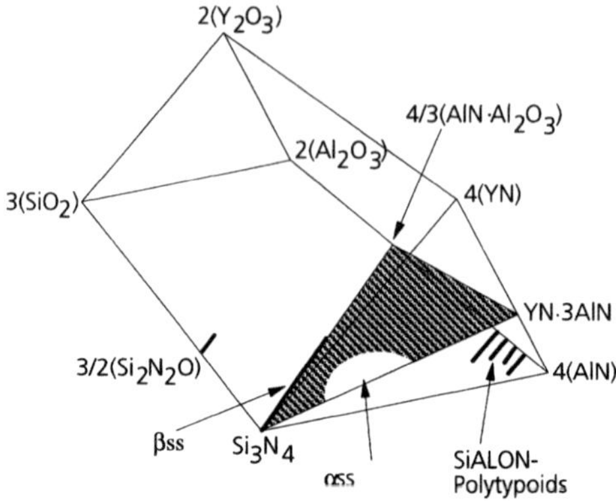

Fig. 9. The Jänecke prism Si₃N₄-4(AlN)-4(YN)-2(Y₂O₃)-2(Al₂O₃)-3(SiO₂) with the αss plane (schematic)

3.3
α-Si₃N₄ Solid Solutions (αss)

In the subsystem Si₃N₄-YN · 3AlN-4/3(AlN · Al₂O₃) a stable α-Si₃N₄ solid solution exists, representing the αss group of Si₃N₄ ceramics. In Fig. 10 an enlarged section of the αss plane illustrates the homogeneity ranges for Nd, Y and Yb.

The α-structure exhibits void positions with seven N atoms as nearest neighbours (Sect. 2). These positions can be partially occupied by ions with an ionic radius of about 0.1 nm, causing a stabilisation of the metastable α-modification (Fig. 11). The cations M which can be incorporated in the α-structure are Li^+, Mg^{2+}, Ca^{2+}, Y^{3+} and lanthanoides with $Z \geq 60$ [123, 169, 187]. Additionally Si^{4+} and N^{3-} must be replaced by Al^{3+} and O^{2-} to obtain electroneutrality. The formula for the resulting α-solid solutions is

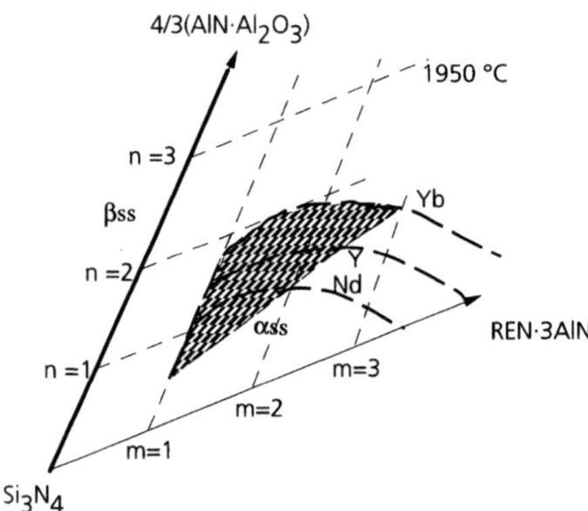

Fig. 10. αss for different rare earth (RE) ions [147]; only the shaded area has been studied. Compare the shaded plan Si₃N₄-YN · 3AlN-4/3(AlN · Al₂O₃) in Fig. 9

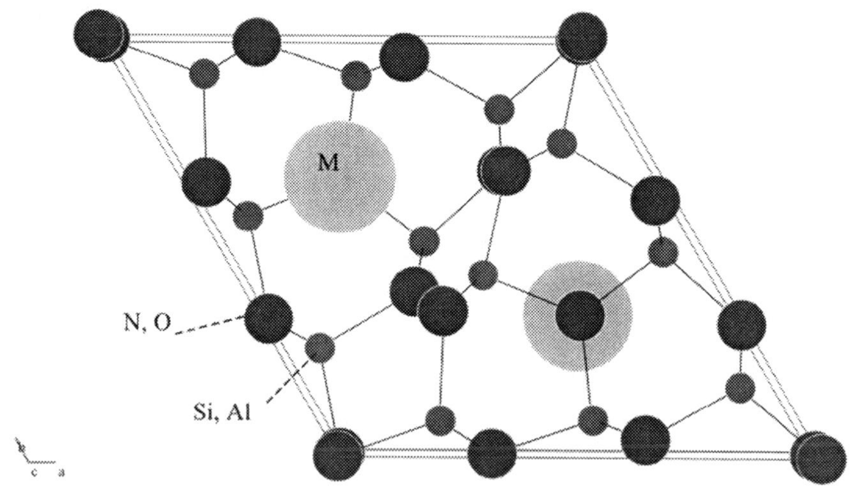

Fig. 11. Projection of the αss crystal structure

$$M_xSi_{12-(m+n)}Al_{(m+n)}O_nN_{16-n} \tag{6}$$

x is the amount of the stabilising cation $M^{v'}$ with the charge v^+, n the amount of oxygen replacing nitrogen. To obtain electroneutrality the amount of Al^{3+} replacing Si^{4+} must be $x \cdot v^+ + n$ in which $x \cdot v^+$ is normally expressed as m.

In addition to Fig. 10 in Fig. 12 some data are given regarding phase relations and stability areas with different rare earth cations. The lowest x-value for trivalent cations is 0.33; for Ca^{2+} the lowest value is 0.3 [123–127, 169, 187]. It should be emphasised that all αss do not include the pure α-Si_3N_4. The atomic positions for different cations M are listed in Table 5, showing that the Si- and N-positions do not differ significantly from that in pure α. This is

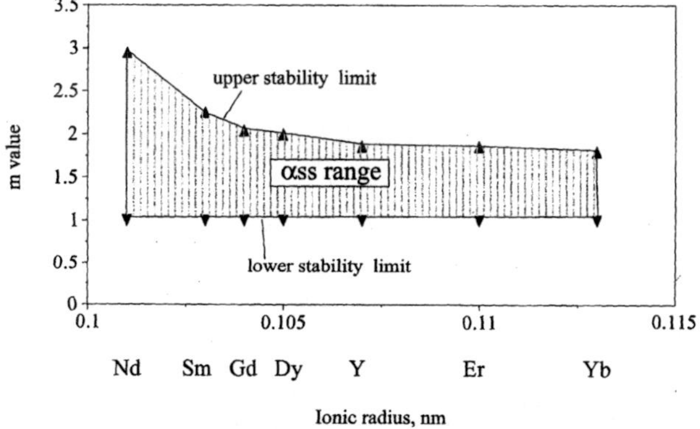

Fig. 12. Range of m values in αss as function of the ionic radii [100]

Table 5. Crystal structure data of different α-solid solutions

Composition		$Nd_xSi_{12-n-m}Al_{m+n}N_{16-n}O_n$	$Y_xSi_{12-n-m}Al_{m+n}N_{16-n}O_n$	$Y_xSi_{12-n-m}Al_{m+n}N_{16-n}O_n$
Parameter		$x = 0.33$; $n = 1.26$; $m = 1$	$x = 0.33$; $n = 1.26$; $m = 1$	$x = 0.5$; $n = 0.75$; $m = 1.5$
Literature		152	152	170
Space group		P31c (No. 159)	P31c (No. 159)	P31c (No. 159)
a, nm		0.78255	0.78167	0.782946
c, nm		0.57008	0.56948	0.570765
RE (Nd, Y)	Occupation factor	0.17	0.170	0.274
(position 2b)	z	0.695	0.691	0.233
(Si, Al)1	x	0.511	0.511	0.5090
(position 6c)	y	0.428	0.427	0.0830
	z	0.660	0.657	0.2338
(Si, Al)2	x	0.170	0.170	0.1717
(position 6c)	y	0.917	0.916	0.2537
	z	0.4498	0.448	−0.0038
(N, O)1	x	0.612	0.611	0.3468
(position 6c)	y	0.956	0.952	−0.0490
	z	0.443	0.446	−0.0092
(N, O)2	x	0.320	0.322	0.3227
(position 6c)	y	0.00634	0.00691	0.3199
	z	0.707	0.704	0.2460
(N, O)3	z	0.107	0.103	0.6462
(position 2b)				
(N, O)4	z	0.452	0.437	0
(position 2a)				

caused by the high rigidity of the $(Si,Al)_3$-$(O,N)_4$ tetrahedron network. Equation (6) demands the homogeneity range of αss on the one plane in the relevant M-Si-N-O-Al systems and no defects in the $(Si,Al)_3(O,N)_4$ tetrahedron framework. This is in full agreement with the experimental results [132, 187].

The maximum cation RE and oxygen solubility in α-solid solutions depends on the ionic radii and increases with decreasing radius of the lanthanoides (Fig. 10 and 12). Solubility limits of the Y-αss phase, $Y_{m/3}Si_{12-(m+n)}Al_{(m+n)}O_nN_{16-n}$ have been determined for the $1.0 \leq m \leq 2.4$ and $n \leq 1.7$ by [155] and as $1.0 \leq m \leq 2.5$ and $0.5 \leq n \leq 1.24$ by [169]. These data differ slightly from more recent ones (Fig. 9b) [147]. Especially the oxide solubilities in the α-solid solutions, e.g., the maximum n values strongly depend on the temperature [147]. Recent investigations based on Rietveld measurements show that, in the Si-N-O-Al-Y system, compositions of αss up to an x value of 0.22 for low n values can be observed at a sintering temperature of 1825 °C [152]. The differences in the experimental results may be caused by difficulties to reach equilibrium (absence of a liquid phase and low diffusion coefficients) and by the interactions with the sintering atmosphere, leading to a shift of the composition. Weight losses of 1–2 at% are quite typical (Sect. 5.3.3). Therefore, the α- and β-lattice parameters calculated by the several formulae are different [127, 155, 169, 171] (Table 4). The reason for the deviations could be incompletely equilibrated samples.

The stability areas of the α-solid solutions depend both on the size of the rare earth ion and the temperature [147]. Increasing the size of the rare earth and decreasing temperature leads to a reduced solubility of Al and O in the αss, e.g., to lower maximum n values [188]. On the other hand, recent results show a more extended αss with more than one stabilising cation, as can expected with only one stabilising cation (different rare earths and Sr or Ca [128]). Even elements like La or Ce, which alone do not form an αss, can be effective as stabiliser together with Ca or Yb [131, 189]. Those multi-cation α-solid solutions offer additional possibilities of variations in processing and properties [154].

The temperature dependence of the stability of the αss could explain why the $\alpha ss \rightarrow \beta ss$ transformation takes place at 1350–1700 K [123, 128, 138, 188] in samples with compositions on the boundary of the αss region or in two-phase $\alpha ss/\beta ss$ mixtures. The transformation is a solution-reprecipitation process which often is kinetically hindered, and the presence of a liquid or low viscosity glassy phase is necessary [188].

4
Si_3N_4 Powders

The majority of Si_3N_4 ceramics are made from powders and therefore depend to a large extent on the quality of the starting powders. Si_3N_4 ceramics of superior quality require well characterized and often extremely pure powders. The powders determine the processing, the sintering behaviour and the subsequent formation of the microstructure which strongly influences many properties of the densified materials.

4.1
Powder Synthesis

Many different synthesis routes for obtaining Si_3N_4 powders are available:

- direct nitridation of silicon [21, 190–196],
- diimide synthesis [191, 197],
- carbothermal nitridation of silica [198–200], alumosilicates [201, 202], or rice husk [203, 204],
- vapour phase synthesis [191, 205, 206],
- plasmachemical synthesis [21, 207–211],
- laser induced reactions [191, 212],
- pyrolyses of silicon organic compounds [213].

All of these synthesis routes result in mixtures of α and β, mainly with a high α to β ratio and are based on four different chemical processes (Table 6).

Presently only the direct nitridation and the diimide synthesis are of commercial importance; the industrial production by gas phase reaction was given up recently.

The **direct nitridation** of Si is carried out at temperatures higher than 1100 °C but below the melting point of Si (1412 °C). This reaction is strongly exothermic and must be precisely controlled to prevent the melting of Si and the subsequent formation of a higher amount of coarse β-crystallites from the melt [191, 216].

Commonly the powders are agglomerated and must be milled [191]. The purity depends on the purity of the raw material and the milling procedure after nitridation. Chemically purified powders usually contain impurities of chlorine and fluorine. Especially the low-cost refractory-grade Si_3N_4 powders produced by fast nitridation have a higher impurity content and are coarser.

An improved nitridation process of Si in a rotary tube furnace allows a better control of the exothermic reaction, avoids overheating and agglomeration [192], is faster and results in smaller crystallites. Those powders have an improved sintering behaviour and after sintering a more favourable microstructure compared to powders produced by a batch process [192]. Recently successful fast nitridation in a fluidised bed reactor was realised in laboratory experiments [194].

The *self-propagating high temperature synthesis* (SHS) uses the heat formation due to the strong exothermic nitridation reaction for self propagation [214, 217, 218]. Due to uncontrolled thermal peaks above the melting point of Si high amounts of coarse β-grains are formed. This is the main reason that the elegant SHS method has not succeeded to a technological extent for production of high quality Si_3N_4 powders.

The *direct nitridation of Si powder by a plasma chemical reaction* takes place at temperatures much higher than the melting point of Si up to 6000 K

[208]. The mean reaction time is very short, therefore only very small α- and β-crystallites (about 20 nm) and amorphous Si_3N_4 are formed. By an additional heat treatment fine α- or β-rich powders can be produced, which allow the production of Si_3N_4 ceramics with mean grain sizes of about 100 nm [219].

The **diimide synthesis** takes place at the interface between liquid ammonia and the organic solvent in which the $SiCl_4$ is dissolved. The product must be washed and calcined to remove the NH_4Cl. At high temperatures (1300–1500 °C) the crystallisation to an α-rich powder takes place [220]. Temperatures above 1500 °C cause an increase of β and grain size [221]. A very fine β-rich powder can be obtained, when the crystallisation of an amorphous Si_3N_4 powder takes place at 1300–1450 °C in presence of an oxide nitride liquid in which the amorphous phase can be dissolved and reprecipitated mainly as β [219].

There are three more possibilities for the powder production by the diimide synthesis [206]. Beside the reaction at interfaces of two liquids the formation of the diimide can be realised by

- bubbling of gaseous $SiCl_4/N_2$ through liquid ammonia at about 0 °C,
- reaction of gaseous ammonia with liquid $SiCl_4$ or $SiCl_4$ solutions in organic solvents at about 0 °C,
- reaction of gaseous NH_3 and $SiCl_4$ [196, 197, 206].

Very similar to the diimide synthesis process is the **vapour phase synthesis**. Different gaseous components {CCl_4 [21, 191], hexamethyldisilazane (HMDS) or other silazanes [210, 222]} react at 800–1400 °C (Table 6) directly to form amorphous Si_3N_4, which has to be crystallised by a further heat treatment. The main starting materials in the gas phase process are $SiCl_4$ and ammonia which react to amorphous Si_3N_4 at 800 °C. Removal of the by-product NH_4Cl, crystallisation of the amorphous powder (1300–1500 °C), and deagglomeration must follow.

Beside the production of pure Si_3N_4, the gas phase processes are used for the preparation of Si_3N_4/SiC composite powders used for nanocomposite materials [210, 222, 223]. For the production of such complex powders different kinds of evaporable Si organic compounds are used, e.g., hexamethyldisilazane.

The reaction in the gas phase can also be achieved in a plasma or laser induced. The result of this process is an amorphous or only partially crystallised powder with a high specific surface area.

The gas phase processes and the diimide process offer a very fine, high purity powder with high sinterability. The main disadvantage of these powders are the poorer shaping behaviour and the higher price in comparison to direct nitrided powders.

The production of Si_3N_4 powders by **carbothermal reduction** has not yet been commercialised. The attempts to produce fine, cheap powders by this method have not been successful [198]. For refractory-grade powders a carbothermal reduction of different minerals to form βss and/or αss (SiAlONs) may play an important role in the future [215].

Table 6. Synthesis of Si_3N_4 powders

Method	Process	Technology/raw materials	Powder features	Literature
1. Direct nitridation	1100–1400 °C $3Si + 2N_2 \rightarrow Si_3N_4$ solid + gaseous	Si-powder; N_2 or N_2/H_2 or N_2/NH_3 gas	α content: <97; variable purity (purification); variable particle size (milling); good shaping behaviour; without purification cheep process	[21, 191, 192, 195, 196]
	>1400 °C $3Si + 2N_2 \rightarrow Si_3N_4$ Solid + gaseous	High temperature self propagating synthesis (SHS)	Coarse powder with high β-Si_3N_4 content	[214]
	<6000 °C $3Si + 4NH_3 \rightarrow Si_3N_4 + 6H_2$ Solid + gaseous	Plasmachemical reaction	Fine particle size, high purity powder; variable degree of crystallinity, variable α/β content	[208]
2. Diimide synthesis	−40–0 °C $SiCl_4 + 6NH_3 \rightarrow Si(NH)_2 + 4NH_4Cl$ 900–1200 °C $3Si(NH)_2 \rightarrow Si_3N_4 + 2NH_3$ 1300–1500 °C Si_3N_4 (amorph) $\rightarrow Si_3N_4$ (cryst.)	Reaction at phase boundaries liquid/liquid or liquid/gas or in the gas phase	High purity powder, high α content fine particle size, special treatment: fine β powders; processing/ shaping more sensitive than for direct nitrided powders	[21, 197, 206]

3. Vapour phase synthesis	300–1600 °C $3SiCl_4 + 4NH_3 \rightarrow Si_3N_4 + 12HCl$ 1300–1500 °C Si_3N_4 (amorph) $\rightarrow Si_3N_4$ 800–1400 °C $3(CH_3)_6Si_2NH + 5NH_3 \rightarrow 2Si_3N_4 + 18CH_4$	Chemical vapour reaction (CVR) HMDS or other evaporable silicon organic compounds Plasma chemical synthesis Laser induced reaction	High purity, high α content, fine particle size, special treatment: fine β-powders, processing/shaping more sensitive than direct nitrided powders Fine, high purity powder, high α content with variable degree of crystallinity; special treatment: fine β powders, difficulties during processing/shaping	[21, 205, 213] [21, 209, 210, 212]
4. Carbothermal nitridation	1450–1600 °C $3SiO_2 + 6C + 2N_2 \rightarrow Si_3N_4 + 6CO$	SiO_2 or kaolin, different carbon qualities; rice husks	Variable purity (purification), variable α/β content; raw material SiO_2: α content: <97; raw material kaolin: high β content	[198, 199, 202, 203, 215]

4.2
Powder Characterisation

Si_3N_4 powders have pronounced differences with respect to crystallinity, α/β ratio, surface area, surface charge, particle shape, particle size, particle size distribution, agglomeration and impurities. This is obvious from Table 6 to some extent and even more from Table 7. As a consequence a precise characterisation of the powders is imperative.

The **chemical composition** with respect to Si and metallic impurities (mainly Fe, Ca, Al) is generally determined by wet chemical methods in combination with standard spectroscopic techniques (AAS, AES, XRF) (Table 8) [224–226]. A precondition is the dissolution of the powder. Typical dissolving processes are fusion with sodium carbonate or mixtures of sodium carbonate and boric acid, with alkaline hydroxides [225, 226] and special acid treatments [225]. A more effective analysis based on optical emission spectroscopy allows the direct analysis of impurities in the solid state and requires no dissolution step [227].

Silicon is a constituent of the main phase as well as of the most common impurities SiC, free Si, Si_2N_2O, SiO_2, $FeSi_x$. Therefore, besides the determination of the total amount of Si, additional methods are needed to determine the amounts of free Si and the other compounds. Quantitative X-ray diffraction (XRD) is often used; with modern XRD methods less than 0.1 wt.% free Si [232] and 0.2 wt.% Si_2N_2O [228] are detectable. The accuracy for SiC is somewhat lower due to the overlapping of the main peak of SiC with β-Si_3N_4. Free Si can be determined also with volumetric methods [225].

Nitrogen can be determined by wet chemical analysis [225, 226] or by inert gas fusion technique at 2700 °C [233, 234].

The total amount of *oxygen* can be analysed by neutron activation [224] or inert gas fusion technique. The inert gas fusion technique with controlled heating rate allows a distinction between surface and bulk oxygen [234–236]. Surface oxygen can be analysed by XPS methods [233, 237]. The oxygen is mainly in a thin SiO_2 surface layer on the Si_3N_4 particles. The total bulk oxygen content consists of different amounts of Si_3N_4 solid solution and Si_2N_2O [228, 234].

Table 7. Typical Si_3N_4 powder characteristics (From different commercial data sheets)

	Direct nitridation			Diimide synthesis	Plasma chemical synthesis
	Refractory grade	Low cost grade	Higher cost grade		
β content, wt%	>25	5–20	5–10	<5	30
Mean particle size, μm	>10	2–5	0.3–0.7	0.3	0.1–0.7
Crystallite size, nm	>200	70–150	70–150	30–70	15–70
Surface area (BET), m²/g	≈1	4–6	10–15	2–13	20–70
Fe impurity, wt%	≤1	<0.1	0.03	<0.01	≤0.01
Oxygen content, wt%	<0.6–2	<1.5	<2	<2	<4

Table 8. Standard methods for the analysis of Si_3N_4 powders

Elements/phases	Method	Literature
α/β ratio	XRD	[224, 228–231]
Si	Wet chemical analysis	[225, 226]
Free Si	XRD,	[228, 232]
	Wet chemical analysis	[225, 226]
O, N, C	Inert gas fusion technique,	[233–235]
	Neutron activation,	[224]
	Wet chemical analysis	[225, 226]
Metallic impurities	Wet chemical analysis,	[224, 225, 227]
	AAS, AES, MS, OES	[224, 227]

The *carbon* content, mostly SiC or WC inclusions, is usually determined by the inert gas fusion technique [204].

Anionic impurities as F^- and Cl^- result mainly from specific powder preparation and purification processes and can be determined by spectroscopic methods [224].

It has to be mentioned that the chemical analysis of the impurities often is insufficient for an adequate powder characterisation because of their very specific influence on the properties of the powder compacts (sintered bodies). On the one hand, particle size and particle distribution of the impurity inclusions are important; for instance it has been shown that the same impurity contents of Fe cause a drastic decrease in strength of the compacts if concentrated in few inclusions with diameters >50 μm, but have nearly no influence as inclusions <5–10 μm [238]. On the other hand, most metallic impurities form silicides which cause a remarkable volume increase [238].

The α-$/\beta$-Si_3N_4 ratio is measured by X-ray diffraction [228–231]. In earlier investigations only some peaks where chosen [229–231]. New techniques based on the Rietveld method use the whole angle range of the XRD diffraction pattern [219, 228]. This results in more accurate data even in textured samples [228]. The amount of amorphous Si_3N_4 also can be analysed by this method using an internal standard [224].

The commercial Si_3N_4 powders mostly consist of 85–98% α; a high amount of α is desired, since it results in materials with high mechanical properties (Sects. 6 and 7).

The **crystallite size** can be determined by XRD [219, 224], which is sensitive for crystallites up to 0.1–0.2 μm; bigger crystallites cannot be determined. This is disadvantageous because the large crystallites have a significant influence on the formation of the microstructure (Sect. 6). The TEM/SEM investigation is very time-consuming due to the large amount of particles necessary to analyse. Additionally, the distinction between α and β is difficult because of the similar diffraction patterns.

The **particle size distribution** affects the shaping processes, green density, the sinterability and the final density. The particles mostly are not single crystallites but contain several crystallites. Therefore, the size measured by XRD is different from that measured by common particle size measurements

such as sedimentation or light scattering techniques. As an integral value of
the particle size often the specific surface area is used, measured by the BET
method [239].

For the **particle size and particle shape** analysis several methods are
employed [224, 240]. The quantitative particle shape analysis is time-
consuming and expensive and rarely used in Si_3N_4 powder characterisation.
Since in regular powders the particles are agglomerated, a main problem of the
measurement is the reproducible deagglomeration, which needs ultrasonic
treatment and a control of the interaction between particles [241]. The particle
interaction can be controlled by adjusting of the particle surface charge with
the help of organic surfactants [241, 242].

For the water-based processing of the powders with submicrometer grain
size, the **surface charge** characterised by the ζ potential is essential for a
homogeneous distribution of the sintering additives (Sect. 5). Therefore the
knowledge of the surface properties is very important [241, 242]. The surface
charge depends significantly on the pH (Fig. 13), on surfactants and on the
surface oxygen, which forms an oxide layer. With increasing surface oxygen
the isoelectric point shifts to low pH, i.e., to the isoelectric point of SiO_2
(Fig. 14). Therefore Si_3N_4 powders mostly will be dispersed or mixed with
sintering additives at high pH values of the dispersion.

5
Consolidation of Si₃N₄ Powders

The compaction of Si_3N_4 powders to parts or components is multi-staged and
in many respects similar to that used in powder metallurgy:

- **Processing of powders** by cleaning, milling, screening, deagglomeration,
  mixing with processing and sintering additives, avoiding environmental
  contaminations, drying.

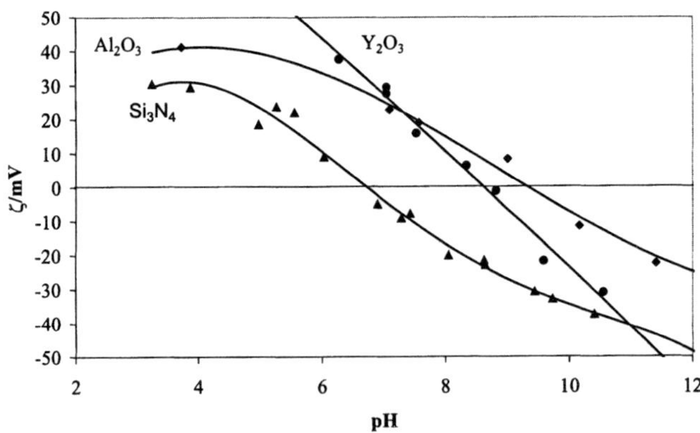

Fig. 13. Dependence of surface charge (ζ-potential) of Si_3N_4, Y_2O_3, Al_2O_3 powders on pH
[241]

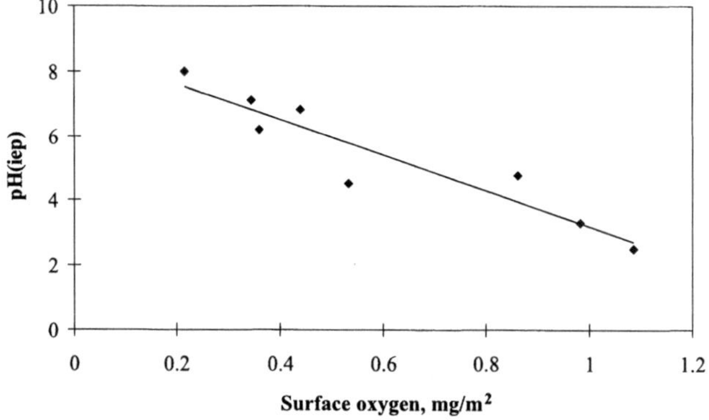

Fig. 14. Dependence of the isoelectric point [pH(iep)] of Si_3N_4 powders as function of surface oxygen [241]

- **Shaping** by axial or isostatic pressing, extrusion, injection moulding, slip casting, tape casting, colloidal methods.
- **Densification** by different sintering methods.
- **Finishing** by cutting, grinding, lapping, polishing, drilling for precision parts with optimised surfaces.
- **Testing and applications** of components (Sect. 10).

Within each stage fracture-causing defects may be introduced which cannot be corrected in the following stages (Table 9) [21, 243]. In all stages there are close correlations with chemistry, chemical engineering as well as solid state and surface chemistry [244]. Probably most efforts in the development of Si_3N_4 ceramics are directed to powder processing and the other consolidation steps, because during these procedures microstructures may be tailored to control properties.

5.1
Processing

In general during powder processing two types of substances are added to the Si_3N_4 powders:

Processing and sintering additives.

5.1.1
Processing Additives

These are dispersants (e.g., amines, fatty acids, polycarboxylic acids), binders (e.g., wax polyacrylate, polyvinyl alcohol), plasticisers (e.g., stearic acids, dibutyl phthalate, polyethylene glycol) and solvents (e.g., water, organics) [244–250]. They are needed to prepare a suitable suspension or slurry for the subsequent shaping step. They must evaporate during a presintering treatment without causing failures (Table 9). Suspensions as well as slurries should have an uniform distribution of the Si_3N_4 particles, adequate viscosity, and no

Table 9. Possible defects occurring in the stages of fabrication of Si_3N_4 ceramics [243]

Powder production (e.g., reaction process)	Powder preparation (e.g., Conditioning)	Powder shaping (e.g., Pressing)	Powder densification (e.g., Sintering)
Unsuitable – Particle size distribution – Mean particle size – Particle shape Deviation from composition Foreign particles Impurities Oversize particles Hard agglomerates	Unsuitable agglomerate size distribution Hard agglomerates Hollow agglomerates Pores in the agglomerates Varying density distribution in the agglomerates Unsuitable viscosity Inhomogeneous distribution of the additives Unstable suspensions Insufficient binder suspension Low content of binder phase Organic inclusions	Bubbles, voids, cracks Micropore clusters Density Inhomogeneities Non-uniform distribution of binder and sintering additions Segregation of binder and sintering additions and of small particles Incomplete removal of the binder Organic inclusions Dust Texturing resulting from particle shape	Large and small pores Micropore clusters Cracks Voids Non-uniform grain growth Exaggerated grains Zones of varying grain size distribution Undesirable grain boundary phases Inclusions Surface roughness

bubbles. To avoid environmental impurities (dust, organic particles, etc.), often clean room or closed loop conditions are required.

The dispersion stability, rheology, and consolidation of numerous aqueous and non-aqueous Si_3N_4 suspensions have been studied extensively [251–257]. Recently a novel class of dispersants for Si_3N_4 powders in non-aqueous media has been designed and its interactions with the powder surface have been characterised systematically on the basis of surface chemistry and fundamentals of colloidal stabilisation [255, 258].

5.1.2
Sintering Additives

They are a precondition for dense Si_3N_4 ceramics. Because of its covalent bonding and low diffusivity Si_3N_4 cannot be densified by common dry sintering. To create a liquid-phase sintering process yielding finally to full densification, additions of sintering aids are necessary. They react with the adhered SiO_2 on the powder surface to a silicate phase, which is molten at the sintering temperature.

Common sintering additives are mixtures of metal oxides (Li_2O, CaO, MgO, SrO, Al_2O_3, RE_2O_3, ZrO_2) or mixtures of oxides with non-oxides like AlN, ZrN, and Mg_3N_2. Often small deviations in composition cause pronounced variations in effectiveness. For instance, it has been shown that the compound $YAlO_3$ has more beneficial effects on sintering than Y_2O_3 + Al_2O_3

mixtures [256]. Mostly, amount and composition of the mixture of additives have been developed empirically. As a first approximation for a good additive the following can be stated: Firstly, under sintering conditions Si_3N_4 should not react with the additives to form a nitride and SiO_2

$$Si_3N_4 + 3\,Me_xO_2 \Leftrightarrow 3\,SiO_2 + 3\,xMeN_y + (2 - 1.5yx)N_2 \tag{7}$$

The free energies of the exchange reactions Eq. (7) $\Delta G_7°$ must be positive, otherwise Si_3N_4 decomposes. Secondly, Si_3N_4 and SiO_2 should not react with the oxides or nitrides to form silicides or metals i.e., the reactions

$$Si_3N_4 + 3\,Me_xO_2 \Leftrightarrow 3\,SiO_2 + 3\,xMe + 2\,N_2 \tag{8}$$

must have large positive free energies ($\Delta G_8°$). These conditions are expressed for several oxides in Fig. 15. Oxides which satisfy both conditions and therefore are suitable as sintering additives lie in the shaded area of Fig. 15 [259].

For instance, TiO_2 reduces to TiN ($\Delta G_7°$ is negative and $\Delta G_8°$ positive) and the d-elements in the 5th to 8th group of the periodic table of elements react to silicides, forming inclusions.

In addition, additives must form eutectics with Si_3N_4 below the sintering temperature. Densification improves with lower eutectic temperature and viscosity of the eutectic melt. Therefore the densification decreases in the direction $MgO/Al_2O_3 > MgO/RE_2O_3$; MgO; $RE_2O_3/Al_2O_3 \gg RE_2O_3$ (RE = Y, Sc, La and lanthanoides).

Amount and composition of the additives are not only of decisive influence on the sintering parameters (temperature, pressure, time, atmosphere), but

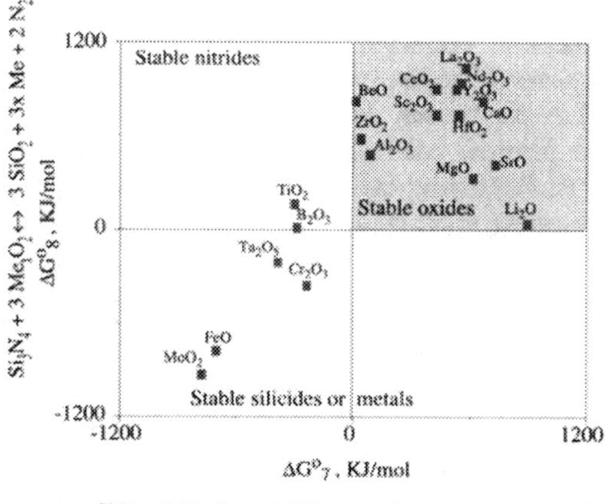

Fig. 15. Plot of $\Delta G_8°$ [Eq. (8)] against $\Delta G_7°$ [Eq. (7)] for some metal oxides at 2000 K. Shaded area indicates stable oxides under sintering conditions, i.e., effective additives (after [259])

also on the resulting phase relations (Sect. 3) and microstructures (Sect. 6), which emphatically determine many properties of Si_3N_4 ceramics.

The additives are conventionally mixed with the Si_3N_4 powders either by ball milling, planetary milling, or attritor milling. It is problematic to achieve full homogeneity in the powder mixtures by these mechanical methods [260, 261]. A clearly improved homogeneity can be reached by incorporating additives into Si_3N_4 powder by using sol-gel techniques [260, 262, 263] or by in situ incorporation by a combustion process [261]. Experiments with combusted Si_3N_4 powders and nanosized sintering aids result in a reduction of sintering temperature compared with conventionally processed Si_3N_4 ceramics [264]. Nevertheless, it should be mentioned that the advantage of perfect homogeneity cannot be fully exploited in the following production steps because of diminishing influences. So far the highest strength values of Si_3N_4 ceramics (up to 1600 MPa) are produced by conventional mixing of additives and Si_3N_4 [245, 246].

5.2
Shaping

All shaping methods (also called "molding") mentioned are well-known in the ceramic industry [244, 247]. Selection criteria depend on economical and technical requirements. Simple shaped parts mainly produced by uniaxial pressing, cold isostatic pressing, casting and especially injection moulding are favoured for components with complex geometry [263].

All shaping methods need processing additives (Sect. 5.1.1): dispersants and plasticisers for suited rheological conditions for slip casting and injection moulding as well as binders for ensuring stability of the powder compacts. Prior to sintering all processing additives have to be burned out quantitatively without inducing defects (Table 9); this procedure is called "dewaxing". Burnout or dewaxing temperatures are comparatively low compared to sintering temperatures to maintain an open porosity; therefore the required times reach from several hours to several days, depending on component thickness and kind of plasticiser. Air is the preferred atmosphere for burning out the binder. To prevent oxidation of the powder the temperature must remain below 750 °C. Following the burnout treatment the green body may be shaped by green or white machining if a precise shape is required before the final densification process.

5.3
Densification

Table 10 summarises all methods for the densification of Si_3N_4 used at present. The resulting Si_3N_4 ceramics classified according to the densification routes are also listed together with several remarks on manufacturing characteristics, properties and applications. For comparison with the sintered qualities, information on reaction bonded silicon nitride ceramics are also included; but will be treated in more detail in Sect. 8.

5.3.1
Densification Methods

The most common densification method is the **gas-pressure sintering** which guarantees better reproducibility and improved properties by only moderate increase in production costs compared to **pressureless sintering**. The first dense Si_3N_4 ceramics were produced by **hot pressing** [13]. Nowadays it is used for the preparation of specimens in materials development or for some applications where components of simple geometries and low quantities are required.

The **hot isostatic pressure (HIP)** methods are high-cost technologies and therefore used only for special applications. A full densification of Si_3N_4 in absence of sintering additives succeeds only with **capsule HIP**; the adherent silica of about 3 vol% on the Si_3N_4 particles acts as sintering aid since it melts during HIPing. The HIP procedure requires encapsulating the Si_3N_4 powder compacts in gas-tight glasses, which soften at the sintering temperature and therefore uniformly transmit the external gas pressure to the powder compact. High densification with very precise retention of shape is a great advantage of capsule HIP, but removing the encapsulation material by mechanical methods or chemical etching is time consuming and costly.

5.3.2
Densification Mechanism

Apart from HIP, all attempts to densify Si_3N_4 powder **without additives** have not been successful. Densification requires sintering additives (Sect. 5.1.2). The role of the additives is to react with Si_3N_4 and its adhered silica to produce a liquid at high temperatures which allows mass transport through solution-reprecipitation to consolidate the solid Si_3N_4 by rearrangement and coalescence in equilibrium with the liquid. Thus the general reaction may be expressed as:

> a) α/β-Si_3N_4 + SiO_2 + additives (starting powder mixture)
>
> $\downarrow$ sintering temperature
>
> b) α/β-Si_3N_4 + liquid of SiO_2, additives and dissolved Si_3N_4
>
> $\downarrow$ cooling   (9)
>
> c) β-Si_3N_4ss + amorphous phase (SiO_2, additives)
>
> $\downarrow$ devitrification temperature
>
> d) β-Si_3N_4ss + secondary phases + amorphous phase

Both modifications α and β are present in different amounts in the starting powder (Table 10); mostly α-rich powders are used. At sintering temperature the metastable α-phase and the subcritical β-particles ($d_{crit.} < 0.5$ µm) dissolve in the liquid and reprecipitate during cooling as β solid solutions on the initial stable β-particles which act as nuclei, while the homogeneous and heterogeneous nucleation can be neglected in most cases (Sect. 6.1).

Table 10. Production technologies and resulting Si_3N_4 ceramics

Densification method	Sintered materials					Reaction bonded materials	
	Low gas pressure sintering	Gas pressure sintering	Hot pressing	Hot isostatic pressing (HIP)		Reaction bonding	Reaction bonding and postsintering
				Presintering	Capsule		
Material	SSN	GPSN	HPSN	HIP-SSN, Sinter-HIP-SN	HIP-SN	RBSN	SRBSN, HIP-SRBSN
Starting powders	Si_3N_4	Si_3N_4	Si_3N_4	Si_3N_4	Si_3N_4	Si	Si + additives
Additive content, vol%	7–20	3–15	2–15	2–15	0–8	0	3–15
Heat treatment	1600–1800 °C nitrogen pressure up to 0.3 MPa	1750–2000 °C nitrogen pressure up to 10 MPa	1500–1800 °C uniaxial pressure in a graphite die	Presintering to closed porosity (similar to SSN); HIP-process: 1750–2000 °C; gas pressure up to 200 MPa	1750–2000 °C gas pressure up to 200 MPa	1200–1450 °C up to 100 h	Nitridation as for RBSN, Sintering as for SSN, GPSN or HIP-SSN
Linear shrinkage %	15–22	15–22	up to 50	15–22	15–22	≈0	Up to 8–14
Main advantages	Low sintering cost, complex shapes possible, continuous sintering process possible	Complex shapes, better reproducibility, better reliability than SSN, lower additive content	Good densification, high reliability	Complex shapes possible, sintering of materials with low sinterability (low additive content), high reliability	Complex shapes possible, materials without additives, high reliability	No shrinkage, low cost raw material, low production costs	low shrinkage, low cost raw material

Main disadvantages	Lower strength than GPSN, good sinterability necessary	Somewhat higher sintering cost than SSN	Only simple shapes, low productivity	High sintering costs	High sintering costs, high costs for encapsulation	Low strength, porosity, low hardness	Time consuming nitridation
Relative density %	95–99	98–100	≈100	≈100	≈100	70–88	95–100
Strength, MPa (20 °C)	500–900	800–1500	800–1500	800–1500	≈500 no additives, 800–1500 with additives	150–350	500–1000
Toughness, MPa m$^{1/2}$	5–8	5–11	5–8	5–11	≈3 no additives, 5–8 with additives	2–4	5–11
Applications	Wear parts, precision parts	Cutting tools, wear parts, ball bearings, seals, engine parts (valves, turbo charger rotors)	Mainly used for evaluation of materials, prototypes with simple geometries, cutting tools	Ball bearings, wear parts, precision parts	Materials with no or very low additive content	Refractories	Same as SSN/GPSN

During densification the green compacts shrink up to 25% depending on green density, starting powder characteristics and amount of additives. Fig. 16 shows typical densification curves of Si_3N_4 ceramics with different amounts of the same additive mixture. Densification starts at 1200–1300 °C when the eutectic melt begins to form, as can be seen in the shrinkage curve dL/L_0 (Fig. 16a) as well as in the densification rate curve dL/dt (Fig. 16b). The first maximum in Fig 16b is caused by the rearrangement of the Si_3N_4 particles and the second by the solution-reprecipitation mechanism, the main densification process. The sintering behaviour improves with increasing additive amount (Fig. 16). However for amounts above 15–20 vol%, gas bubbling occurs caused by an increased formation of gaseous SiO [265, 266].

After cooling, the stable β-phase is embedded in an amorphous phase, from which an oxide nitride can be precipitated during a devitrification heat treatment, which is done just below the eutectic temperature. But even after

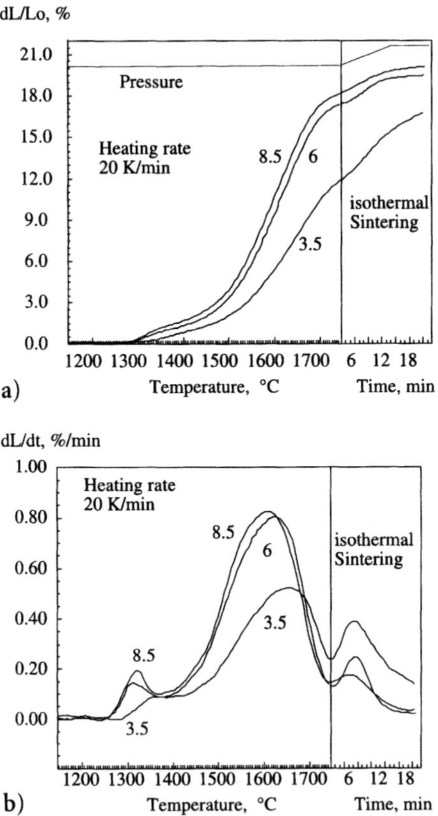

Fig. 16. Typical shrinkage dL/dLo (a) and densification rate dL/dt (b) curves of Si_3N_4 ceramics. Additives: Y_2O_3/Al_2O_3 mixture of weight ratio 2:1 and contents of 3.5, 6.0 and 8.5 vol%

very long annealing times a certain amount of amorphous phase remains (Sect. 6). Depending on the composition of the starting powder mixtures the steps c) and d) in Eq. (9) yield different phase relations according to the relevant phase diagrams (Sect. 3).

Theoretically, single-phase βss ceramics can only be achieved with additives in the systems Si-N-O-Al and Si-N-O-Be in which extended β solid solutions exist (Sect. 3.2). With respect to Eq. (9) in step c) or at the latest in d) homogeneous βss material without other phases emerges. Nevertheless, small deviations of the compositions from the theoretical one, which is normally the case, result in the formation of small amounts of residual grain boundaries. Also, in the theoretically monophase materials thin amorphous grain boundary films are stable, and indeed, in HIPed (2273 K, Ar pressure 180 MPa) high purity Si-N-O-Al powders glass films at the grain boundaries are obtained (Sect. 6.1.4).

In compositions of the starting powder mixture relevant to Si_3N_4 ceramics of low β solid solution (Sect. 3.2.2), a polyphase material according to Eq. (9d) is originated. For compositions resulting in stabilised αss ceramics (Sect. 3.3), theoretically a single-phase material can also be produced by tailoring the composition. But usually in these materials thin grain boundary films also remain. The reaction steps b)–d) in Eq. (9) can be applied in the analogous way for these materials. An α solid solution is existent because the stabilising elements are dissolved in α mostly during heating between step a) and b) in Eq. (9).

Finally, one should take note of the fact that in all Si_3N_4 ceramics a residual, continuous glassy phase remains between the crystalline phases, even after prolonged annealing time at temperatures above 1300 K. This amorphous phase is neglected in all phase diagrams (Sect. 3), yet strongly affects the properties. Therefore much effort has been made to characterise the amorphous phase which will be treated in more detail in Sect. 6.1.4.3.

In conclusion, it can be stated that Si_3N_4 ceramics are polyphased materials including mainly βss, αss, secondary phases (mainly oxide nitrides, in ceramic literature generally called oxynitrides) and an amorphous phase, all having characteristic morphologies and can be arranged in a manifold of microstructures (Sect. 6).

5.3.3
Influence of Densification Parameters

Besides the influence of the starting powders and the sintering additives, temperature, time, pressure, and atmosphere are important parameters which must be taken into account.

Of course, increasing sintering temperatures and time enhance densification, but above optimum values they cause unfavourable changes of the microstructure (Sect. 6). In general, increasing pressure during hot pressing and HIPing causes increased densification rate, refinement and for hot pressing, an anisotropic orientation of the microstructure as well.

5.3.3.1
Decomposition

Because of the decomposition of Si_3N_4, the sintering temperature is limited, and the vaporisation of the melt formed by reaction of the additives with the adherent silica begins. The decomposition can cause a shift of composition in the near surface area of the sample during sintering [232, 267–269, 272–274]. To suppress decomposition, high nitrogen pressure in the sintering atmosphere is required. This is achieved in the powder bed technique and encapsulation technique. The formation of SiO according to

$$3\,SiO_2(l) + Si_3N_4 \Leftrightarrow 6\,SiO + 2\,N_2 \tag{10}$$

is the main reason of decomposition during sintering [232, 267, 270–272].

Overheating or a flowing nitrogen atmosphere may cause additional decomposition by evaporation [274]:

$$Si_3N_4 \Leftrightarrow 3\,Si + 2\,N_2 \tag{11}$$

In a non-flowing atmosphere the evaporation becomes more pronounced when the Si_3N_4 nitrogen partial pressure reaches the equilibrium partial pressure of Eq. (11). The reactions (10) and (11) explain the dependence of the weight losses on the ratio of the mass or volume of the Si_3N_4 samples to the gas volume in the furnace. The colour changes of the samples during sintering are also caused by this ratio [232] (Sect. 7.4).

Amount and composition of the sintering additives may also be changed by interaction with the atmosphere. For instance, Al_2O_3 can be partially reduced to AlN which will be dissolved in the β-Si_3N_4 lattice (Sect. 3) according Eq. (12):

$$0.5\,z\,Al_2O_3 + (2 - 0.25\,z)Si_3N_4 \Leftrightarrow Si_{6-z}Al_zN_{8-z}O_z + 0.25\,z\,SiO_2 \tag{12}$$

In this case Al_2O_3 acts like a buffer, the more SiO_2 evaporates, the more Al dissolves in the silicon nitride grains. The amount of Al in the Si_3N_4 solid solutions depends, besides the sintering parameters, on the Al_2O_3 activity in the oxide nitride liquid, i.e., on the ratio of Al_2O_3 to the other additives [269]. Strong reducing atmospheres and high weight losses cause an intensive reduction of Al_2O_3 and SiO_2 which may even result in α-SiAlON formation on the surface [232, 274].

Rare earth oxides and Y_2O_3 are more stable than the other additives [267]; no change in concentration of yttria or lanthanoide oxides has been found experimentally, even during long sintering times [232, 268, 272, 274].

MgO, commonly used as sintering additive, is unstable and evaporates according to:

$$6\,MgO + Si_3N_4 \Leftrightarrow 6\,Mg(g) + 3\,SiO_2 + 2\,N_2 \tag{13}$$

The alkaline and other alkaline earth oxides behave similarly.

5.3.3.2
Influence of Carbon

Carbon or graphite heating elements and insulating materials, as commonly used in sintering devices, may have a deleterious influence on the densification process. Carbon or SiC can cause a mass loss in the sintered compact during sintering because of the reduction of SiO_2 according to:

$$3\,SiO_2(l) + 6\,C + 2\,N_2 \Leftrightarrow Si_3N_4 + 6\,CO \text{ (high } N_2 \text{ pressure)} \tag{14}$$

$$SiO_2(l) + 2\,SiC + 2\,N2 \Leftrightarrow Si_3N_4 + 2\,CO \text{ (low } N_2 \text{ pressure)} \tag{15}$$

At higher nitrogen pressure, C is in equilibrium with Si_3N_4, at lower nitrogen pressures, SiC. Thus at 1800 °C the conversion of SiC to C and Si_3N_4 takes place at a nitrogen pressure of 6 MPa. There is a pronounced dependence of the stability ranges of the species on nitrogen pressures and temperature [275]. The formation of CO is the reason for the mass loss, which should increase with increasing N_2 pressure, assuming equilibrium. But under usual sintering conditions a reduced weight loss with increasing nitrogen pressure is observed [232, 272, 274]. This indicates a non-equilibrium of the CO partial pressure. The formation of CO seems to be kinetically determined by transport processes involving SiO, which is reduced by increasing nitrogen pressure [232, 272]. The weight loss can be minimised by reduction of the H_2O and O_2 content in the atmosphere because both react with graphite heating elements and form CO. An SiC surface layer on the carbon parts in the sintering devices reduces CO formation for kinetic reasons, and that is why aged carbon crucibles are advantageous as experience shows.

Another way to reduce the decomposition due to the interaction with the atmosphere in a carbon containing furnace is sintering under CO partial pressure [268]. If the CO partial pressure in the furnace reaches the equilibrium pressure (Eqs. 14, 15), CO is no longer formed and consequently, the weight loss is reduced. But if the pressure is higher than the CO equilibrium pressure, additional SiO_2 and C are formed (up to 0.6% C) [268]. The presence of carbon causes structural defects (pores, concentration gradients, dark coloured spots), and therefore has to be suppressed or minimised for a reproducible sintering process. Additionally, high CO partial pressure results in difficulties in temperature control, and the life of the heating elements is reduced.

5.3.3.3
Influence of Powder Bed

Si_3N_4 ceramics have been, and sometimes still are, sintered in a powder bed to prevent the decomposition and interaction with carbon and/or graphite elements of the sintering devices [270]. The powder beds usually consist of Si_3N_4/BN mixtures and occasionally sintering additives are added [276, 277]. The powder bed may react with carbon parts in the furnace [274]. A critical evaluation of the relevant results must take into account the dependence of SiC and C stability on the nitrogen pressure and compositional changes of the gas phase.

Sintering in a powder bed has several disadvantages compared with normal sintering (low thermal conductivity, temperature gradients, increased scatter of sample shape and properties in large furnaces, and more complicated handling during packing and removal of the powder). Therefore sintering without a powder bed is increasingly favoured and has gained acceptance because of improved sinterability of Si_3N_4 powders, better sintering furnaces and by better understanding of the sintering processes. Control of interactions between atmosphere and material is essential for reproducible sintering. Nearly complete densification and weight losses below 1% can be achieved without powder bed.

5.3.4
Densified Materials

A classification of dense Si_3N_4 ceramics follows from the densification techniques (Table 10):

Sintered Si_3N_4	SSN
Gas-pressure sintered Si_3N_4	GPSN
Hot pressed Si_3N_4	HPSN
Presintered hot isostatically pressed Si_3N_4	HIP-SSN (Sinter-HIP SN)
Encapsulated hot isostatically pressed Si_3N_4	HIP-SN

Silicon nitride ceramics are not merely only one material but several classes of materials. All of them are multiphased, i.e., they exhibit a heterogeneous microstructure which has formed during sintering (Sect. 6). Therefore in all classes a large variety of properties is predominant and as a consequence also a large variety of potential applications (Sect. 10). Often little variations in the powders and the processing parameters cause remarkable changes in the microstructure which have a pronounced effect on properties (Sects. 6 and 7).

5.4
Finishing

In general, components of Si_3N_4 ceramics have to fulfil high performances. Therefore, precise and careful finishing steps are indispensable. Usually cutting, grinding, drilling, and polishing are used but they are time consuming and expensive. Because of the great hardness of Si_3N_4 (Table 1) mainly diamond tools are necessary. Ultrasonic erosion, laser cutting and spark erosion are very promising and increasingly used tools for finishing components of Si_3N_4 ceramics [278–280].

5.5
Precursor-Derived Si$_3$N$_4$ Ceramics

During the past years the pyrolysis of precursors, mainly Si- and N-containing polymers, has gained interest as a production method and to some degree as an alternative to the described powder route [213]. Both monophase Si$_3$N$_4$ and Si$_3$N$_4$ composites with compositions in the systems Si-C-N and Si-B-C-N have been prepared free of additives. After pyrolysis precursor-derived ceramics are amorphous and require an additional heat treatment >1300 °C for crystallisation [Structure and Bonding, Vol. 101 (2002) p 137]. So far, the precursor line is restricted to small and thin components because a short diffusion path is required for the release of the gaseous species developed during pyrolysis. The gas causes uncontrolled cracks, pores and shrinkage with the consequence of pronounced deterioration of the mechanical properties. To overcome these disadvantages the "active filler concept" has been developed [281]. It is based on the idea of using additional fillers (for instance, Ti, Cr, Mo) which react with the gaseous products and reduce them. In addition, the volume change caused by the solid reaction products (mainly nitrides, carbides and carbonitrides) allows a better control of shrinkage.

6
Microstructures

In addition to the crystal structure, the microstructure has a significant influence on the characteristic properties of a material. Type, amount, arrangement, size, shape and orientation of the various phases all contribute to the actual microstructure of a material which thus results from the combination of all phases and the defects they contain. Such defects are vacancies, dislocations, grain boundaries, pores and cracks. The microstructure is an important domain within the science of materials. The higher the requirements of a material, the more stringent are the requirements on its microstructure, i.e., the more accurately must its microstructure be established. The aim is to create a microstructure specifically designed to produce a given property profile. The terms "microstructural engineering" and "microstructural design" are the key words used to describe this problem.

The amount, distribution, size, morphology of the α- and β-particles, secondary and amorphous phases of Si$_3$N$_4$ ceramics are decisive factors for their quality and reliability. In the following, mainly sintered and gas pressure sintered Si$_3$N$_4$ ceramics will be discussed because of their extraordinary economical interest and because they show all the microstructural features also present in the other Si$_3$N$_4$ ceramics.

6.1
Development of Microstructures in βss Ceramics

Silicon nitride ceramics are mostly produced from α-rich Si$_3$N$_4$ powders which transform during sintering by a solution reprecipitation process to β solid solutions (βss) according to the reaction sequence in Eq. (9). Under conventional sintering conditions the low βss and, but to a lesser amount, the extended βss, form elongated needle-like grains with ratios of grain length to thickness (aspect ratios) comparable to whiskers. This is the basis for the concept of an in-situ whisker reinforcement of Si_3N_4 ceramics [282].

The microstructural development is controlled mainly by the Si_3N_4 starting powders, the additives and the sintering parameters.

6.1.1
Influence of Starting Powder

Model experiments with oxide nitride glasses have revealed that the growth of β-grains starts from initial β-particles present during liquid phase sintering [Eq. (9), step b] [283, 284]. Until now no homogeneous nucleation was observed in the oxide nitride liquid [283, 285]. Only sometimes heterogeneous nucleation of βss on α-particles was detected during sintering of α-rich starting powders, obviously due to the similar crystal structures [286, 287]. Epitaxial growth of each modification on the other was found in βss and αss as well, because of the lower differences of the lattice parameters and the reduced nucleation energy compared to the pure α- and β-modifications [288]. In general the starting powders contain β-particles (≥ 2 vol%). During sintering, grain size and morphology of the growing β-particles are strongly influenced by the number and size of the initial β-Si_3N_4 particles. A small number results in a large interparticle distance between β-grains, and therefore the grains are able to grow during liquid phase sintering [Eq. (9), step b] without significant impingement by other grains. In case of β-rich powders (>30%) with particle sizes of about 1 μm, the driving force for the solution reprecipitation process is much smaller. The resulting microstructure is more or less fine grained and characterised by equiaxed grains [282, 289]. Fig. 17a and **b** show two microstructures of different starting powders densified under the same conditions. Similar or even finer microstructures can be obtained from very fine β-powders (grain size $\leq$50–100 nm measured by XRD) by hot pressing and further heat treatment (Fig. 17c and **d**) or gas pressure sintering [219, 290]. Obviously the growth rate of βss in pure β-powders does not differ so much from that in α-rich powders in case of fine powders, but decreases noticeably in coarser β-rich powders [291].

Nevertheless, the microstructure cannot be directly correlated to the initial β-content in the starting powder. The experimentally determined particle density in sintered samples indicates that only a part of the initial β particles are able to grow [283]. The number of growing β-particles depends on a critical particle diameter d_{crit} (Fig. 18). Particles below d_{crit} will dissolve in the oxide nitride liquid during phase transformation and reprecipitate on the overcritical β-particles according to an anisotropic Ostwald ripening process [284, 292, 293].

Besides the sintering parameters (Sect. 6.1.3), d_{crit} is determined mainly by three parameters [290, 294, 295]. First, the size distribution of α- and β-particles due to a higher stability of large α-particles compared to small

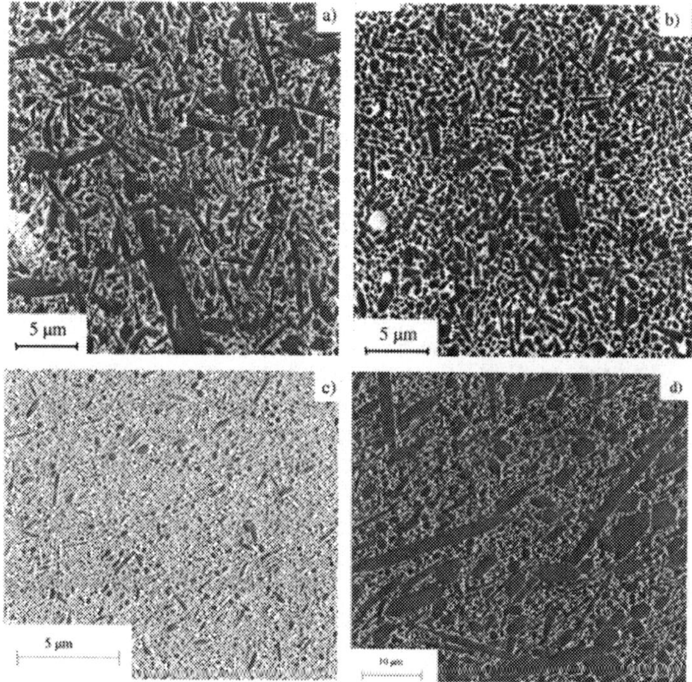

Fig. 17a–d. Microstructures of βss ceramics (plasma etched). **a** from α-rich starting powder. **b, c, d** from β-rich starting powders; **a** and **b** pressureless sintered; **c** hot pressed, and **d** sample **c** additionally heat treated

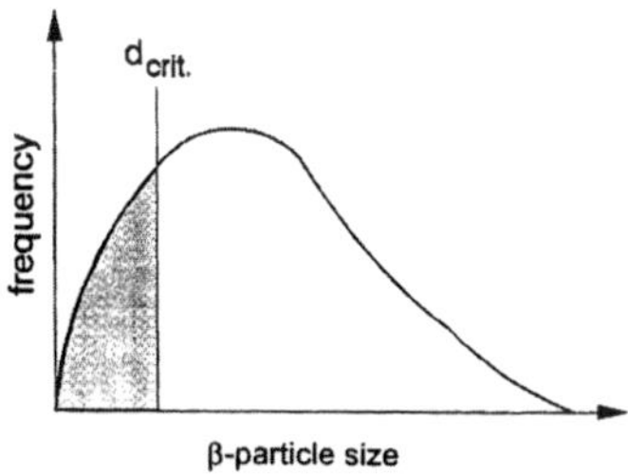

Fig. 18. Schematic of β particle size distribution of a Si_3N_4 starting powder

β-particles in the early stage of sintering. A high volume fraction of large α-particles cause an increase in d_{crit}. Second, the nitride solubility in the siliceous melt which is determined by the additive composition. A higher nitride solubility shifts d_{crit} to higher values. Third, the volume fraction of β. If the powder does not contain enough α, a higher amount of smaller β-particles will dissolve to saturate the siliceous melt, and d_{crit} increases (e.g. in cases of β-rich starting powders).

Furthermore, detailed experiments showed that the microstructural evolution is controlled by grain impingement. Theoretical consideration of the Si_3N_4 crystal structure (Sect. 2) revealed a higher stability of the prism planes of the hexagonal shaped grains compared to basal planes. However, the growth rate of basal planes is much higher than that of prism planes, resulting in needle like grain morphology. The difference in growth rates can be related to the Si_3N_4 crystal structure and is attributed to an energetically more favourable attachment of a surface nucleus on a basal plane. The (001) planes are "atomically rough", and the growth mechanism is diffusion controlled and much faster [283]. The surface energy of the basal plane, i.e., the stability in the melt depends on the thickness of the grains [15, 285, 296]. The (100) plane is "atomically flat" and the growth velocity is determined by the formation of surface nuclei [297, 298]. The surface energy of the prism plane increases with decreasing aspect ratio and thickness. Consequently thick elongated grains grow anisotropically during the densification process and may result in exaggerated grain growth at temperatures above 1800–1850 °C [296, 299]. This fact is used in the production of seeded materials (Sect. 9.3).

Based on the assumption of a high stability of prism planes and a higher growth rate of the basal planes, three microstructure controlling cases of grain impingement can be derived (Fig. 19). The growth of a basal plane will be stopped, if it hits a more stable prism plane (A). The prism plan can cause a diameter reduction of a growing basal plane (B). If the diameter of a basal plane is large enough it will grow around smaller grains and dissolve them (C). It is important to point out that cases (A) and (B) will stabilise the microstructure against grain growth. However, at higher temperatures and long sintering times exaggerated grain growth of particles with a large initial diameter occur by the mechanism schematically shown in Fig. 19C.

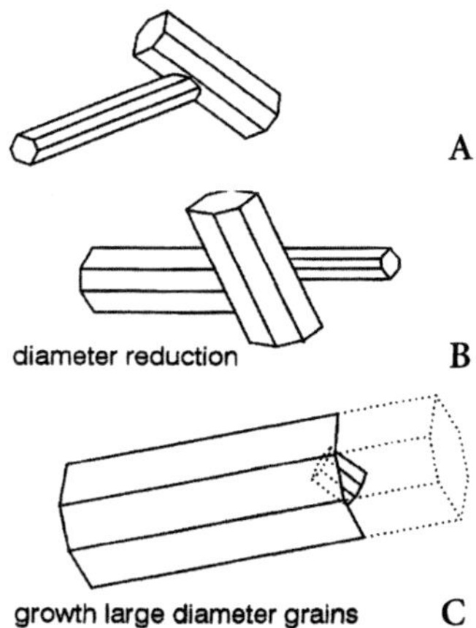

Fig. 19. Schematic of the steric hindrance of grain growth

6.1.2
Influence of Sintering Additives

The influence of the densification aids is pronounced, as can be seen for example from Fig. 20 [300]. The results presented are based on studies by using an additive combination of Al_2O_3 and RE_2O_3 (RE = La, Nd, Gd, and Yb). The additive compositions were calculated in consideration of the different nitride solubility to keep the volume fraction (V_{liq} = 0.15) of liquid-forming additives constant. The plots in Fig. 20 indicate a finer-grained microstructure with high aspect ratios for the La-containing sample compared to the material densified with Yb_2O_3.

The ratio of additive combinations also influences the microstructural evolution. For instance, with decreasing ratio of Y_2O_3/Al_2O_3 the microstructure becomes finer and the aspect ratio lower [301, 302]. MgO as well as CaO additives accelerate the grain growth and increase the aspect ratio [303, 304].

6.1.3
Influence of Sintering Conditions

The heating rate can influence the microstructural development. Fast heating leads to a coarser microstructure due to the dissolution of fine β-Si_3N_4 particles during the heating period in the oxide nitride melt [290, 305]. On the other hand, low heating rates in the interval 1500–1700 °C intensify an exaggerated grain growth [306]. This seems to be caused by the reduced steric hindrance of growing particles and growth processes involving gas phase

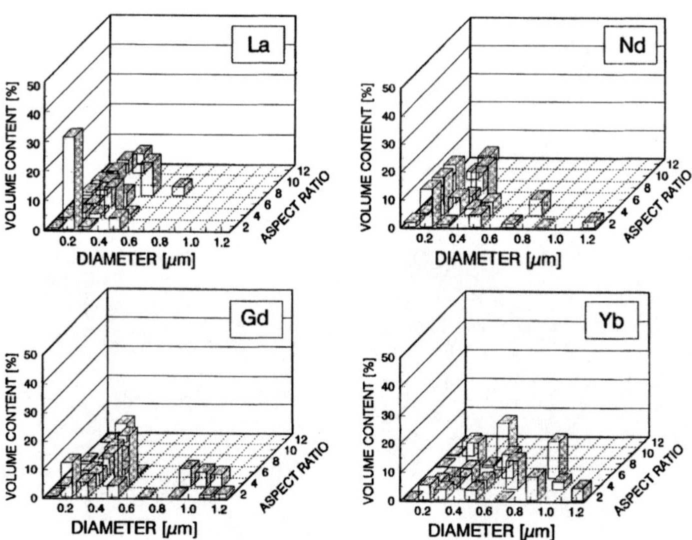

Fig. 20. Grain diameter and aspect ratio distribution for Si_3N_4 ceramics densified with various sintering additives

transport processes. Similar exaggerated grain growth was observed in a two-step sintering procedure, lowering the temperature of the first step below 1750 °C [307].

Sintering temperature and time are other important parameters, not only for densification (Sect. 5.3.4) but also for the microstructural development [308]. After complete phase transformation and partial devitrification according to Eq. (9) step c, grain growth starts by dissolution of smaller β-grains, as concluded from Fig. 21. An increasing aspect ratio with increasing sintering temperatures up to 1950 °C and a decrease with further temperature increase is observed [309]. The observed development of a needle- or rod-like microstructure is caused by the particle size distribution at the beginning of the grain growth stage. Broad particle size distributions lead to a faster grain growth than very narrow ones. Some grains with a large initial diameter can grow in the length direction with a minor steric hindrance up to a length of 100–200 µm [310, 311]. Fig. 21 shows clearly the different types of steric hindrance explained schematically in Fig. 19. Big elongated grains can also form by coalescence during the final sintering [286].

6.1.4
Microstructural Features

From the micrographs shown in Figs. 17 and 21 it is evident the βss grains are embedded in an amorphous or partially crystallised matrix; their amount

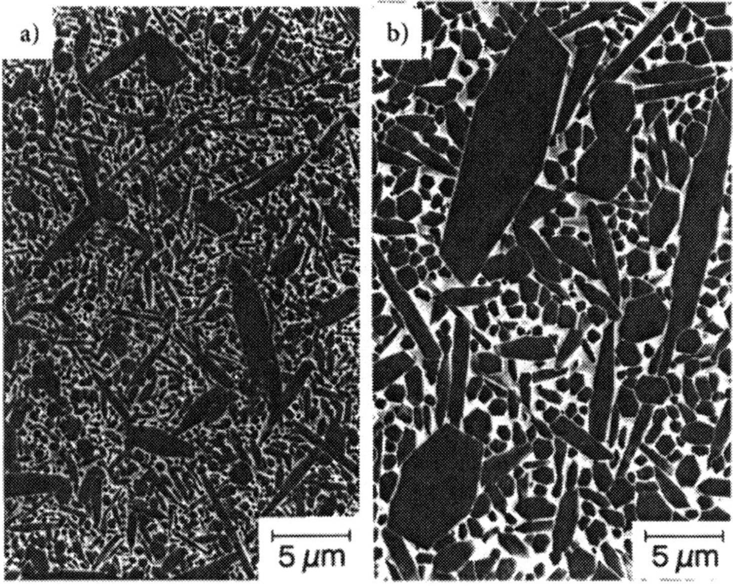

Fig. 21. Microstructure of gas pressure sintered Si_3N_4 after 35 min at 1835 °C (a) and 360 min at 1900 °C (b), respectively

and chemistry are determined by the sintering aids. According to the corresponding phase diagrams the amorphous phase is not in an equilibrium condition. In order to permit a controlled crystallisation, postsintering heat treatments are necessary for devitrification. Nevertheless, the devitrification of the glassy phase is not completed by this treatment. Detailed high-resolution transmission electron microscopy (HRTM) studies reveal that heat treating only partially crystallises the residual amorphous pockets, as shown in Fig. 22a. Complete crystallisation of these pockets cannot be attained because residual glass remains at the tip of each triple junction. Furthermore, homophase (between two Si_3N_4 grains) as well as heterophase (between Si_3N_4 and a crystalline secondary phase) two-grain junctions always remain amorphous. Generally it can be stated that the microstructure of Si_3N_4 ceramics is characterised by three features: the Si_3N_4ss grains, the secondary crystalline phases and the amorphous grain boundary films, as shown schematically in Fig. 22b.

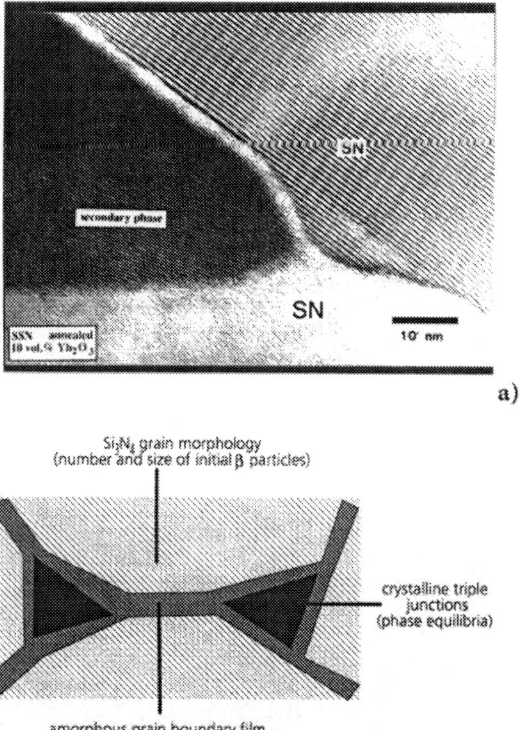

Fig. 22a, b. Microstructural features in Si_3N_4 ceramics. **a** HRTM of a pressureless sintered Si_3N_4 ceramic with 10 vol% Yb_2O_3. Grain boundary film between the Si_3N_4 grains (SN) as well as between the secondary phase and SN grains. **b** schematic of grain boundary region after post-sintering/devitrification treatment. Complete devitrification cannot be achieved

6.1.4.1
βss Grains

As mentioned before, the morphology is given by the number and size of the initial β particles, sintering temperatures and time. An important goal is to have a uniform or bimodal grain size distribution of grains with high aspect ratio in a fine-grained matrix of elongated grains [295, 299, 312–315].

Due to the low diffusion coefficient in the βss grains a concentration gradient of Al arises, which clearly outline the nuclei of the large βss grains (Fig. 23) [283]. In large grains different subgrain boundaries and inclusions of the grain boundary phase have been observed by TEM [286].

6.1.4.2
Secondary Phase

The composition of the crystalline secondary phases can be predicted by the phase diagram (Sect. 3). According to the phase diagrams the resulting Si_3N_4 ceramics would either contain two phases, i.e., compositions reaching equilibrium on a tie line (Si_3N_4 + sec. phase), or three phases, i.e., composition reaching equilibrium in one of the compatibility triangles containing Si_3N_4 as an end-member (Si_3N_4 + sec. phase I + sec. phase II). When further components are included, the phase equilibrium must be viewed in compositional space, and therefore, dense Si_3N_4 ceramics may contain more than three equilibrium phases. Thus number and type of equilibrium secondary phases depend on the amount of each of the starting constituents (Si_3N_4, SiO_2, additives, impurities) and their phase relations. Generally the secondary phases are oxide nitrides; a few examples are given in Table 11. From analytical electron microscopy investigations it has been shown that the secondary phases are highly enriched with rare earth and Al_2O_3 additives at the triple grain junctions while they are less concentrated at the two grain junctions (Fig. 22b). It is also found that the secondary phases are not uniformly distributed within the triple junctions where they are more

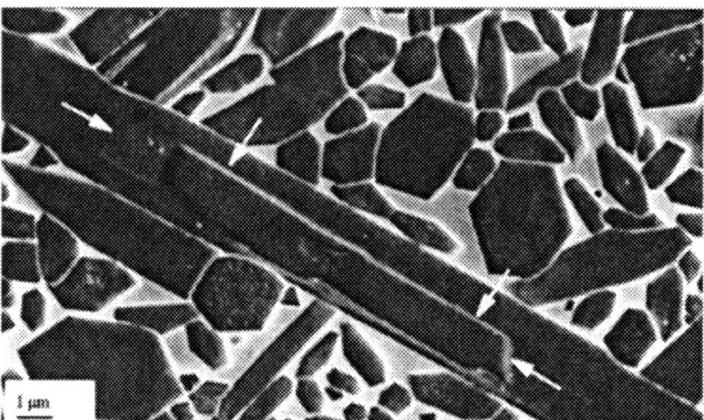

Fig. 23. βss needle grown on β particle as nucleus

Table 11. Grain boundary phases in Si_3N_4 ceramics [101]

Compound	Structure type	La	Ce	Pr	Nd	Pm	Sm	Eu	Gd	Tb	Y	Dy	Ho	Er	Tm	Yb
$RE_4Al_2O_9$	$RE_4Al_2O_9$						X	X	X	X	X	X	X	X	X	X
$REAlO_3$	Perovskite	X	X	X	X	X	X	X								
$RE_3Al_5O_{12}$	Garnet								X	X	X	X	X	X	X	X
$REAl_{11}O_{18}$	$PbFe_{12}O_{19}$	X	X	X	X	X	X	X								
$REAl_{12}O_{18}N$	$PbFe_{12}O_{19}$	X	X	X	X	X	X	X								
$REAl_{11+x}O_{18}N_x$ ($0 \leq x \leq 1$)	$PbFe_{12}O_{19}$	X	X	X	X	X	X	X								
RE_2AlO_3N	K_2NiF_4	X	X	X	X	X	X	X								
RE_2SiO_5	Orthosilicate	X	X	X	X	X	X	X	X	X	X	X	X	X	X	X
$RE_2Si_2O_7$	Pyrosilicate	X	X	X	X	X	X	X	X	X	X	X	X	X	X	X
$RE_{9,33}(SiO_4)_6O_2$	Apatite	X	X	X	X	X	X	X	X							
$RE_{10}(SiO_4)_6N_2$	Apatite	X	X	X	X	X	X	X	?	?	X					
$RE_2Si_3O_3N_4$	Melilite	X	X	X	X	X	X	X	X	X	X	X	X	X	X	X
$RE_2Si_{3-x}Al_xO_{3+x}N_{4-x}$	Melilite		X	X	X	X	X	X	X	X	X	X				
$RESiO_2N$	Wollastonite		X	X	X	X	X	X	?	?	X					
$RE_4Si_2O_7N_2$	Wöhlerite	X	X	X	X	X	X	X	X	X	X	X	X	X	X	X
$RE_4Si_{2-x}Al_xO_{7+x}N_{2-x}$ ($0 \leq x \leq 2$)	Wöhlerite								?	?	X	X	X	X	X	X
$RE_3AL_{3+x}Si_{3-x}O_{12+x}N_{2-x}$	U-Phase ($RE_3Ga_5GeO_{14}$)	X	X	X	X	X	X	X	X	X	X	X	X			
$RE_4Si_9Al_5O_{30}$ N	W-Phase (Latiumite)	X	X	X	X											

X: stable phase.
Empty cells: phase not stable.
?: no data.

concentrated at the central regions, as compared to the surrounding regions at the triple pockets [316].

6.1.4.3
Amorphous Phase

The amorphous phase remains as a stable film at the grain boundaries [317–321]. Its thickness depends on the temperature [312, 313], the additive composition [314, 322, 323] and on the impurities of the starting powder [324]. It varies between 0.5 and 1.5 nm [325] and is constant for a certain additive composition and independent of the amount of additives [317, 318]. An increasing additive content causes only an enlargement of the three- and four-grain junctions, but no change in film thickness of the two grain junctions. Small variations in the chemistry of the amorphous film may result from the βss grain growth [316, 326].

The experimental observations confirm theoretical considerations which predict a stable grain boundary film in Si_3N_4 ceramics through a balance of the attractive van der Waals forces and the various repulsive interactions across the grain boundary [318]. In a more sophisticated treatment of this problem the diffuse interface approach has been used to describe a flat interface between the two coexisting phases [327]. The main assumption of this approach is that the free energy vs. concentration between the two phases can be represented by a continuous function of the Gibb's energy, as if a miscibility gap existed in the system considered. In other words, an interaction energy term exists, describing equilibrium conditions for the amorphous film and its finite thickness. Recently, molecular dynamic calculations have proved that atomic Si-N and Si-O bonds contribute to this energy [320]. However, the apparent stability of the glassy phase that would be expected to crystallise during subsequent heat treatment (devitrifiation) demands an additional explanation. It has been suggested that the crystallisation of the intergranular films can be inhibited due to stresses caused by volume changes during crystallisation [319]. On the other hand, a model experiment has proved the stability of the amorphous intergranular films, even in a geometry free of residual stresses and capillary effects [328]. Thus the question arises whether the existence of amorphous phases in the grain boundary can be explained by the thermodynamic equilibrium in the ceramic system considered; in other words, whether or not the amorphous phase must be taken into account in the phase relations and therefore must be included in phase diagrams. Further arguments for this possibility are given from a molecular dynamic simulation in silicon as a model for covalent materials, identifying a thermodynamic criterion for the existence of thermochemical stable disordered intergranular films based on the relative energies of the atoms in the grain boundaries and in the bulk amorphous phase [329]. A subsequent step is a thermodynamic treatment of boundary phase admitting computations of realistic phase structure in liquid phase sintered ceramics [330]. Thus the additional interaction responsible for getting the thin intergranular film have been considered as a constraint for the thermodynamic equilibrium of the system.

The method of parallel tangent construction developed for representation of equilibrium between bulk and liquid phase at sintering temperature can be used for the description of the stability of the grain boundary films. To examine the true equilibrium of the system the Gibb's energy G^* ($\Delta\mu$) computed assuming the amorphous phase as the boundary interface, has to be compared with the Gibb's energy G^0 ($\Delta\mu$) calculated in the same way, but taking for the grain boundary interface the crystalline equilibrium phase conforming to the conventional phase diagram. A necessary condition for the amorphous grain boundary phase in equilibrium with the matrix is a positive value of the thermodynamic driving force $\Delta G(\Delta\mu)$:

$$\Delta G(\Delta\mu) = (G^0(\Delta\mu) - G^*(\Delta\mu)) > 0 \tag{16}$$

This approach has been applied for Si-Al-O-N ceramics. The results can explain the existence of the amorphous phase in the grain boundaries (Fig. 24), depending on the overall composition of the system ([equ% O] = 16, 12, 8 and [equ% Al] = 10.6) and on the strength of interaction ($\Delta\mu$). In a strict sense it must be concluded that the conventional phase diagrams of Si_3N_4 ceramics are incomplete representations of materials containing an amorphous intergranular phase.

6.2
Development of Microstructures In αss Ceramics

Compared to βss, the investigations on αss microstructures are less numerous and intensive, but many analogies exist. Similar to βss ceramics the αss ceramics (Sect. 3.3) are produced by liquid-phase sintering involving the steps explained in Eq. (9). The starting powder consists of α-rich Si_3N_4 ($\alpha > 80\%$), the additives (mainly AlN, Al_2O_3) and a compound with an appropriate cation which is solvable in the α structure (e.g. Li^+, Mg^{2+}, Ca^{2+}, Y^{3+}, Nd^{3+}-Lu^{3+}). This reaction starts at temperatures above 1450 °C [331]. The amount of liquid

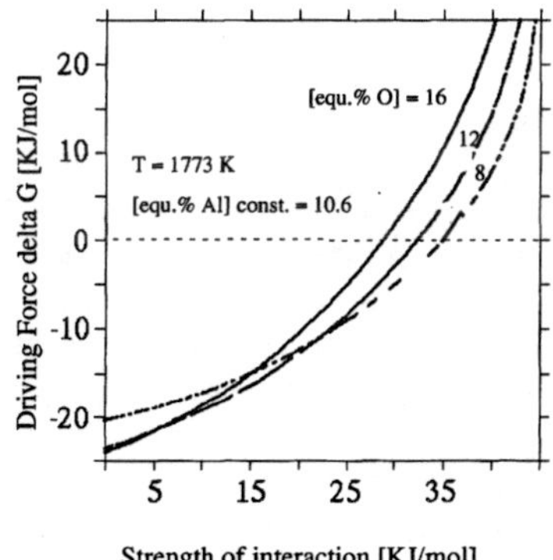

Fig. 24. Driving force for an amorphous grain boundary phase in equilibrium with βss as a function of composition in the system Si-N-O-Al and strength of interaction [330]

available for densification is quickly reduced due to the formation of αss solid solutions. Pure αss ceramics with low amounts of stabilising cations are therefore difficult to densify completely; improved densification can be realised at higher amounts [Eq. (6); high n and m values] [332, 333] or in two-phase ceramics αss/βss [334]. Whilst compositions with low amounts of stabilising cations form microstructures with fine equiaxed grains in the densified ceramics, the compositions with relatively high n and m values [127] or mixed cation compounds (rare earths with Sr or Ca) lead to microstructures with elongated αss grains. Elongated grain growth can be obtained by using β-Si_3N_4 as starting powder [333] as well as α-Si_3N_4 [121, 332, 335, 336]. The size of the elongated grains is larger than those in βss ceramics, indicating the importance of grain growth. However, the reason for the anisotropic grain growth of αss has not been as intensively studied as in βss but the mechanism seems to be quite similar [337]. Starting with α-Si_3N_4 powders the αss grains form epitaxially on the existing α particles. Also, epitaxial growth of αss on βss has been observed [288]; nucleation in the oxide nitride liquid does not seem necessary [340].

A precondition for growing of elongated αss grains is enough liquid at sintering temperatures. Besides the mentioned larger amounts of the cation stabiliser this can be achieved by sintering at higher temperatures ($\geq$1900 °C), reducing the formation rate of the αss by using β powder [337, 338], special heating rates [337], seeds [338, 339] or by an additional stable liquid [340, 341]. Recently it has been shown that elongated αss ceramics with low n and m values can be produced by gas pressure sintering, using an appropriate liquid phase [340]. The growth is also influenced by the nature of the stabilising cations [337].

β-Si_3N_4 as starting powder retards the reaction rate, and therefore the liquid phase exists up to higher temperatures, allowing a more pronounced elongated grain growth [342]. Coarse β starting powder (3 μm) can lead to a reduced formation of the αss for kinetic reasons and increase the grain size of the resulting αss [339].

6.3
Development of Microstructures in αss/βss Ceramics

Since the relevant systems include two phase regions between α and β solid solutions a production of ceramics with microstructures of both phases is possible [343]. Depending on the composition of the α-rich starting powder mixture, the formation of the phase assemblage of the αss/βss ceramic can be realised by all the densification methods described in Sect. 5.3.1. In general, the equilibrium in the two phase region αss + βss is approached in about 1 to 2 h at temperatures between 1700 °C and 1800 °C; times increase with decreasing temperatures. Amount and ratio of the two phases, as well as the lattice parameters, vary with composition, in agreement with the equilibrium conditions in the relevant phase diagrams (Sect. 3). The aim of those compositions is to combine the advantages of αss, namely high hardness with the toughness of βss [334, 336]. Typical microstructures of mixed αss + βss ceramics consist of needle-like βss grains in a matrix of equiaxed αss grains with some amount of secondary phase at the grain boundaries [344].

6.4
Characterisation of Microstructure

For reliable determination of the microstructure, a perfect preparation of the specimens is an absolute necessity [345].

Materials are three-dimensionally arranged. Thus, the two-dimensional information which is gained from the examination of microsections or thin foils, the types of specimen usually used for materialographic investigations, does not truly reflect the whole structure of a material. This becomes clear from the photomontage (Fig. 25a). It shows a cube of silicon nitride ceramic in

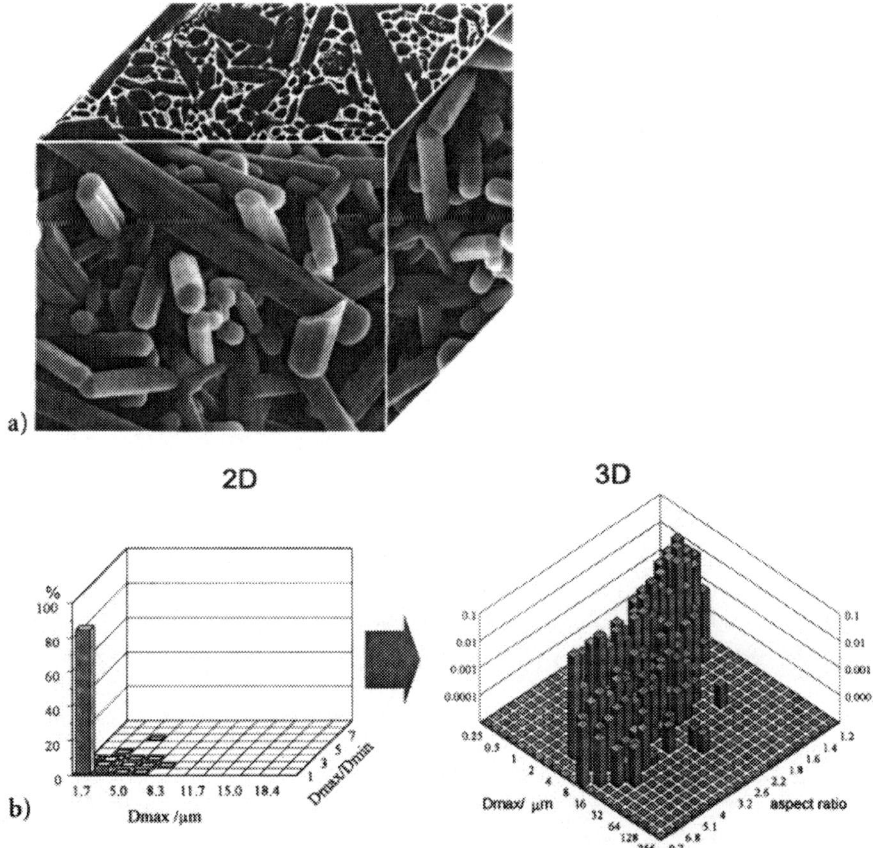

Fig. 25a, b. Stereological analysis of microstructures of Si_3N_4 ceramics. **a** photo montage of an Si_3N_4 ceramic superimposed upon a pile of Si_3N_4 crystallites. **b** the statistic extrapolation of a microstructure with a high proportion of elongated grains

which the matrix phase has been dissolved away leaving behind a pile of Si_3N_4 grains. The upper surface of the cube represents a normal two-dimensional section. It is obvious that the microstructural morphology shown in the section, and in particular the grain size (D) and degree of elongation (aspect ratio) of the grains which significantly affect the fracture behaviour of the material, are not correctly quantified. The quantitative characterisation of the two microstructural features, grain size and aspect ratio, becomes possible only if the visible surface in the two-dimensional section can be extrapolated to reveal the spatial shape of the grains. The transformation shown in Fig. 25b is achieved using a stereographic-based variation of the principles used in microstructural modelling [346, 347]. The histogram shown in the left part of the diagram has been constructed solely from two-dimensional measurements on the specimen. Accordingly more than 80% of the grains would have an almost equiaxial shape. A statistical extrapolation of the grain population, however, reveals a very different size-aspect ratio distribution, as can be seen in the right-hand part of the diagram. This is because in this type of microstructural modelling, the many different ways in which the individual grains can be intersected (Fig. 26) are taken into account, according to their size, aspect ratio and orientation. Often two-dimensional pictures show only minor deviations between the microstructures of materials with pronounced differences in properties, if one ignores the single elongated grain which has quite by chance been sectioned completely along its longitudinal axis. In comparison, the three-dimensional evaluation reveals quite significant differences between the materials.

This basic approach to the special interpretation shows that there are correlations which cannot be studied satisfactorily using two-dimensional sections. However, they also show that there are very efficient methods to extrapolate three-dimensional realities from two-dimensional measurements.

6.5
Microstructure/Property Relations

As a consequence of the large diversity of microstructures which can adjusted by microstructural design, Si_3N_4 ceramics are a whole class of materials with an inherent large variety of properties and therefore a large variety of potential applications (Sect. 10). Different qualities depend on amount and distribution of the microstructural features (Sect. 6.1.4). Often small variations have severe

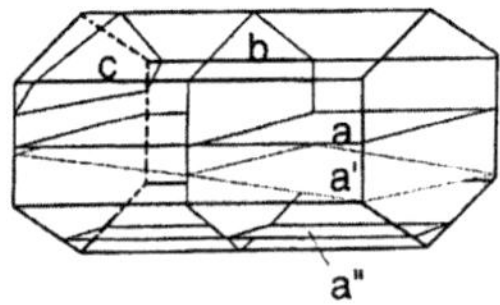

Fig. 26. Some selected cases for planar sections through the hexagonal prism. Generally, triangles up to octagons have to be considered

consequences with respect to the property profile. A more general overview on the microstructure/property relation is in Table 12.

Materials with a high room temperature strength exhibit a fine-grained, elongated microstructure, while materials with a high fracture toughness are more coarse-grained [348]. In both cases, a weak grain boundary phase is required to introduce transgranular fracture. Since all βss grains are completely wetted by the grain boundary phase, the interface strength is determined by the additive composition [349]. Nevertheless, a contradiction arises between the development of high strength, high toughness Si_3N_4 ceramics and high temperature resistant materials because the grain boundary phase is responsible for the excellent properties at low temperatures, but limits the properties at temperatures above its softening point.

There is a significant reduction in strength when microstructures consisting of a broad grain diameter distribution are generated. When a fine equiaxed microstructure is generated, both the fracture strength and the fracture resistance are reduced [33, 350].

In Young's modulus and hardness of extended βss decreases linearly with increasing number of replacing ions z [Eq. (5)]. The decrease was ascribed to the lattice softening which was confirmed by Raman spectroscopy; at high

Table 12. Overview on microstructure/property relations of Si_3N_4 ceramics

Property	Microstructural features	
	Grain size/shape	Grain boundary phase
High strength up to 1000 °C	Fine grained/needle-like grains	Median additive content
High strength at T > 1200 °C	Fine grained/needle-like grains	Al_2O_3 free, special compositions
High fracture toughness	Large, needle-like grains, or large, needle-like grains in a fine matrix	Low Al_2O_3 and SiO_2 content
High hardness	Fine grained or αss/βss or αss materials	Low additive content
High fatigue strength (cyclic mechanical load)	Fine grained/needle-like grains	Low additive content
High heat conductivity	Large grains, low amounts of impurities and defects and solid solutions	No components solvable in the Si_3N_4 crystal lattice, (e.g., Al, Be), no impurities
High creep resistance	Large, needle like grains, composites with SiC, or refractory silicides	Al_2O_3 free with special compositions or no sintering additives
High oxidation resistance at T > 1200 °C	Large, needle-like grains, composites with SiC, or refractory silicides	Al_2O_3 free with special compositions or no sintering additives
High corrosion resistance		Special compositions depending on the corrosive media
Good wear behaviour	Fine-grained microstructure	Homogeneous distribution

temperatures, creep rates were significantly enhanced with increasing z values [351]. In general, the creep resistance decreases with increasing amounts and decreasing viscosity of the intergranular amorphous phase, but depends also on size and distribution of Si_3N_4 grains and the oxidation behaviour [352–355]. The oxidation behaviour strongly depends on the additives [352, 356, 357].

The hardness of αss is significantly higher than that of βss (Table 1) [332–337, 340] and, in case of elongated grain morphology, the fracture toughness can be significantly improved by crack deflection, crack bridging and grain pull-out mechanism up to K_{IC} values of 6.3 MPa$^{1/2}$ [358].

The rising fracture resistance with the extension of the crack (R-curve behaviour [349, 359]) is more expressed in coarse-grained materials [360]. The R-curve behaviour also has a significant contribution with respect to thermal shock behaviour; during thermal shock a continuous strength degradation is observed in materials with a pronounced R-curve behaviour, in contrast to a catastrophic failure of fine-grained materials with a minor R-curve behaviour [360, 361]. In Si_3N_4 ceramics with gradient microstructures compositions, microstructures and properties change gradually from the hard αss with spherical grains on the surface to the tough and strong βss with elongated grains in the core, a promising combination of microstructures for wear applications [362].

The microstructure/property relations are treated in more detail in Sect. 7.

7
Properties

Si_3N_4 ceramics have a bunch variety of interesting properties. They are light, have good mechanical and thermomechanical behaviour, they are compatible with metals, and they are wear and corrosion resistant. The unique inherent properties of the Si_3N_4 modifications (Sect. 2), the reasonable alloying behaviour (Sect. 3) and the variability in the production processes (Sects. 4 and 5) open a wide range for tailoring microstructures (Sect. 6) and remarkable property combinations.

7.1
Physical Properties

The **electrical resistivity** of Si_3N_4 is $>10^{14}$ Ωcm at room temperature and $>10^6$ Ωcm at 1200 °C [363]. The **electrical conductivity** of βss only slightly increases with increasing degree of alloying substitution z according to Eq. (5) at 700 °C (2×10^{-9} $(\Omega$cm$)^{-1}$ for z = 1.5; 2×10^{-7} $(\Omega$cm$)^{-1}$ for z = 3.2) [364, 365]. The conductivity is ionic at temperatures above 900–1000 °C and electronic by impurities below [364, 365]. At temperatures below 800 °C the activation energy is around 1 eV and 1.8–2 eV greater around 1000 °C. Therefore Si_3N_4 or βss ceramics are used as electronic insulators. Li-containing αss has an electrical conductivity at high temperatures which is similar to that of solid electrolytes, e.g., β-alumina; the conductivity of the αss composition $Li_2Si_9Al_3ON_{15}$ is 9×10^{-6} $(\Omega cm)^{-1}$ and the activation energy of 0.93 eV is low. An electroconductive βss material (ionic conduction) was developed by designing the grain boundary phase [366]. Dense sintered and hot pressed Si_3N_4 ceramic exhibit an electric break down strength >200 kV cm^{-1} [367]. The low dielectric losses tan δ and dielectric strength ε (6–8.0 at 1 MHz) in a broad frequency range is advantageous for radar windows [368, 369], but disadvantageous for microwave sintering [370].

The **specific heat** of Si_3N_4 ceramics is in the temperature range 293 up to 1200 K $[C_p$ (293 K) = 0.67 KJ (K kg)$^{-1}]$ nearly independent of the composition of the additives. The isobaric specific heat values agree well with the isochoric specific heat calculated by Debye's theory. Also the Dulong Petit's rule can applied as an approximation of the C_v values [25 J(K mol)$^{-1}$] at temperatures >1100 K [371]. From the C_p values at around 100 K the amount of the amorphous grain boundary phase can be calculated [371].

Thermal diffusivity a and thermal conductivity λ strongly depend on the composition (Fig. 27). Materials containing a high residual α content show a very low **thermal diffusivity** (88 vol% α; a = 5 mm^2 s^{-1}) whereas the same composition after complete α/β transformation has 13 mm^2 s^{-1} at RT. αss ceramics also show low thermal diffusivity (Fig. 27). The intrinsic anisotropic thermal diffusivity inside the individual grains is without connection with the macroscopic diffusivity, as a consequence of the dispersion of grain orientations in the ceramics [375].

The **thermal conductivity** of βss depends strongly on the amount of Al impurities and/or sintering additives because the incorporation of Al and O in the β structure reduces the thermal conductivity of the grains due to the reduced free path of phonons. At RT the thermal conductivity initially increases with increasing grain size before reaching constant values [372]; it depends on internal defects in the grains (dislocations, Al and O impurities, point defects) [373]. Al-free sintering additives are a precondition for the production of Si_3N_4 ceramics with high thermal conductivity. With increasing sintering time the thermal conductivity increases, because of grain growth and

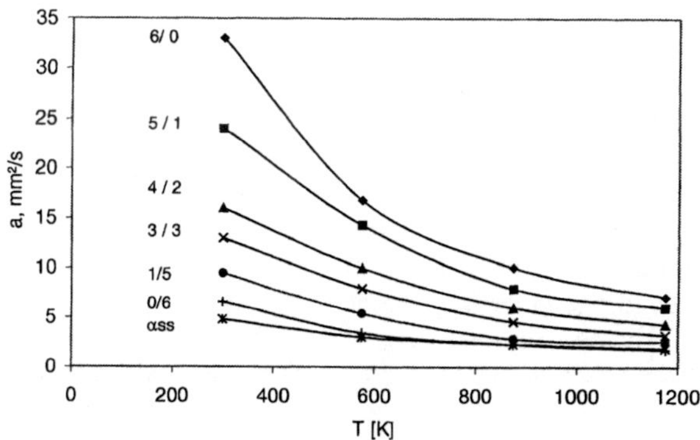

Fig. 27. Thermal diffusivity versus temperature of βss with different Y_2O_3/Al_2O_3 ratio from 6/0 to 0/6 (mol/mol) [371] and αss (n = 1.4, m = 1) [379]

defect healing [374]. The thermal conductivity of β-Si_3N_4 ceramics strongly decreases with increasing grain boundary film thickness [372]. In hot pressed ceramics with different rare earth oxide additives the thermal conductivity increases in the order La [31.6 W(mK)$^{-1}$], Nd [81.6 W(mK)$^{-1}$], Gd [100.7 W(mK)$^{-1}$], Y, Yb [115 W(mK)$^{-1}$], because of the decreasing amount of oxygen dissolved in the β grains [374, 376]. The highest values for the thermal conductivity of isotropic Si_3N_4 ceramics are around 100 W(mK)$^{-1}$, for ceramics with anisotropic grain orientations values of 160 W(mK)$^{-1}$ were measured [377] (Sects. 2 and 9.3).

The **linear thermal expansion coefficient** α of the βss materials depends only slightly on the additive composition [301]. Between RT and 1000 °C values between 3.1 up to 3.6×10^{-6} K^{-1} are common; they increase slightly with increasing temperatures. For β-Si_3N_4 ceramics containing 6 wt% Y_2O_3 and 4 wt% Al_2O_3 $\alpha = 2.8 \times 10^{-6}$ K^{-1} between RT and 200 °C, and between RT and 500 °C $\alpha = 3.2 \times 10^{-6}$ K^{-1}.

Compared to βss, the linear thermal expansion coefficient α(RT–1000) for αss is higher; depending on the cation RE in $RE_{0.6}Si_{9.3}Al_{2.7}O_{0.9}N_{15.1}$ values between 3.8×10^{-6} K^{-1} (RE = Er; Nd) and 3.7×10^{-6} K^{-1} (RE = Y) and for the Ca containing αss 3.9×10^{-6} K^{-1} were measured [378].

7.2
Mechanical Properties

Especially strength and fracture toughness are essential for structural applications. Therefore, great efforts are made to optimise all preconditions for a reproducible production of reliable Si_3N_4 ceramics with high strength and toughness [19].

The **strength** σ of a brittle material is proportional to the fracture toughness K_{IC} and indirectly proportional to the square root of the highest defect size $\sqrt{a}$ in the loaded volume: $\sigma = Y \cdot K_{IC}/\sqrt{a}$ (Y is a geometry factor). For optimisation a must be reduced and K_{IC} increased. Whilst the reduction of a is possible by optimising all the processing steps (Sect. 5), an improvement of K_{IC} is mainly possible by microstructural engineering (Sect. 6).

Stresses arising during application may cause defect growth (subcritical crack growth) which at room temperature only is activated for stresses close to the fracture stresses, but at elevated temperatures above the softening point of the glassy grain boundary phase, they are more pronounced. In addition, softening of the grain boundary phase (T_g) is the reason for deformation processes at high temperatures (creep). The typical behaviour of Si_3N_4 ceramics under static load is shown schematically in a fracture map (Fig. 28). At temperatures up to the softening point of the glassy phase (for oxide nitride glasses formed by common sintering additives between 800–1000 °C) the materials behave in a brittle fashion with low subcritical crack growth and high oxidation resistance (Sect. 7.2.1). At temperatures higher than the transformation temperature, creep, creep fracture, subcritical crack growth and damage due to oxidation become the dominating processes (Sect. 7.2.2). For the measurement of mechanical properties several methods are available [380–383].

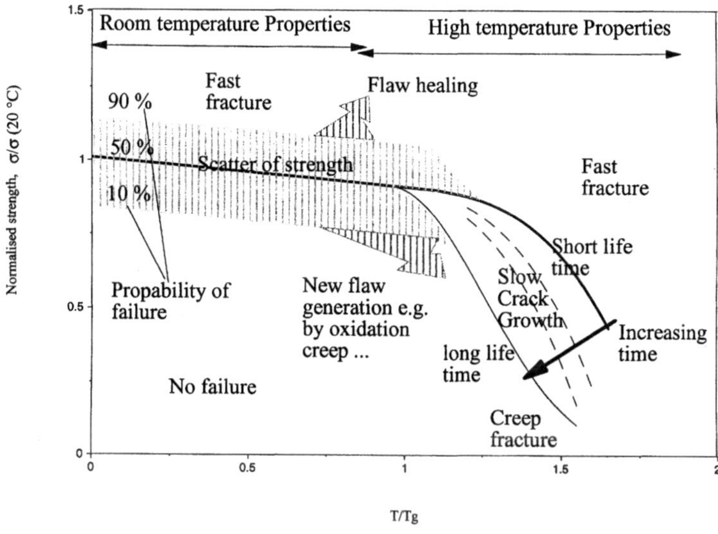

Fig. 28. Fracture map of Si_3N_4 ceramics schematic [382]

7.2.1
Room Temperature Properties

The **strength** of commercial Si_3N_4 ceramics are in the range of 800 to 1400 MPa, depending on defects like pores, cracks, inclusions such as iron silicides or agglomerates of sintering additives. In materials with higher strength or high fracture toughness, unusual elongated grains can be strength-determining defects. So far, the highest strength for isotropic Si_3N_4 ceramics is 2000 MPa measured by three point bending tests [246]. Recently a strength of 2100 MPa was measured for materials with anisotropic grain orientation produced by super plastic forging [384]. Materials with strengths between 1400 to 1500 MPa usually have defect sizes of about 10 µm [245, 310]. Therefore their grain size should be below 10 µm, because larger grains can act as strength reducing defects.

For strength levels above 1000 MPa a special surface finishing is necessary because surface defects are strength limiting. Compression stresses in the surface reduce the effective tensile stresses acting on surface defects and thus increase the strength. In contrast tensile surface stresses reduce the strength. Internal stresses, different sizes and orientations of surface defects may reduce the strength by more than some hundred MPa [385]. A careful reproducible finishing of the ceramics for strength testing and application at high loads is absolutely necessary [385]. In the as-sintered state the strengths of the Si_3N_4 ceramics are usually lower than after removing of the surface layer which is decomposed during sintering. 750 MPa is measured for materials in the as-sintered state with minimised decomposition reactions during sintering [386]. Porous Si_3N_4 ceramics with oriented β whiskers and high strength have a high strain to failure relation at RT due to reduced elastic constants [387].

The **fracture toughness** varies in a wide range from 3 to 12 MPa m$^{1/2}$, this is on one hand connected with variations in the microstructure and on the other, by different methods of determination giving slightly different values [380, 381, 383]. The fracture toughness depends strongly on the microstructure. Two main factors influence the fracture toughness: grain shape and size, and the composition of the grain boundary phase.

The comparatively high fracture toughness of Si_3N_4 ceramics in comparison to other ceramic materials is related to the toughening mechanisms which are similar to those in whisker reinforced composite materials: grain bridging, pull-out, crack deflection and grain branching around large, elongated grains [359, 388–390]. Due to these mechanisms, the fracture toughness increases with increasing volume and the square root of the mean grain thickness of the elongated grains (grains with aspect ratio > 4) [304, 349, 359, 388, 390–393].

The dominant toughening mechanism depends on the grain thickness of the elongated grains. Elastic bridging and pull-out were observed for thin, needle-like grains (thickness < 1 μm). Crack deflection was mainly observed for thick, elongated grains (thickness > 1 μm), whereas grain bridging was detected independent of the grain size [388]. These mechanisms are also responsible for a pronounced rising fracture resistance with crack extension (R-curve behaviour) [303, 304, 390, 391].

The toughening mechanisms can only operate when the dominant fracture mode is intergranular. The ratio of transgranular to intergranular fracture depends on the relative strengths of the grain boundaries and the grains. For a material with a high toughness, the grain boundary must be weak in comparison to the grains (Fig. 29) [33, 389]. Materials with large, needle-like grains in the matrix (in-situ-reinforced) can be produced by prolonged sintering times, special additive combinations or by seeding the materials with β-Si_3N_4 whiskers. Seeding leads to a more narrow size distribution of the large grains and allows one to achieve textured materials with large, oriented β-Si_3N_4 needles [377, 394] (Sect. 9.3). A similar anisotropy was observed in hot pressed materials with low amounts of sintering additives [395] or in superplastic forged materials [384]. Perpendicular to the hot press direction a higher strength and fracture toughness is observed than in the parallel direction.

Recent investigations show that fine-grained materials have a higher toughness than coarse-grained materials in the small crack region (crack length < 30 μm) but lower toughness in the large crack region (crack length > 50 μm) [396–398]. This can be important for the applications in which high local stresses exist (e.g., for ball bearings).

The strength of the grain boundary is connected with two different mechanisms: local residual stresses [399] and special chemical interactions between the grain boundary phase and the Si_3N_4 grains [33, 389]. Generally, the amorphous or partially crystallised grain boundary phases have thermal expansion coefficients different from Si_3N_4. When the thermal expansion coefficient of the grain boundary phase is higher than that of the Si_3N_4 grains, the grain boundary phase is under tensile stresses and the fraction of intergranular fracture is high. This is the case for nearly all Si_3N_4 ceramics. In

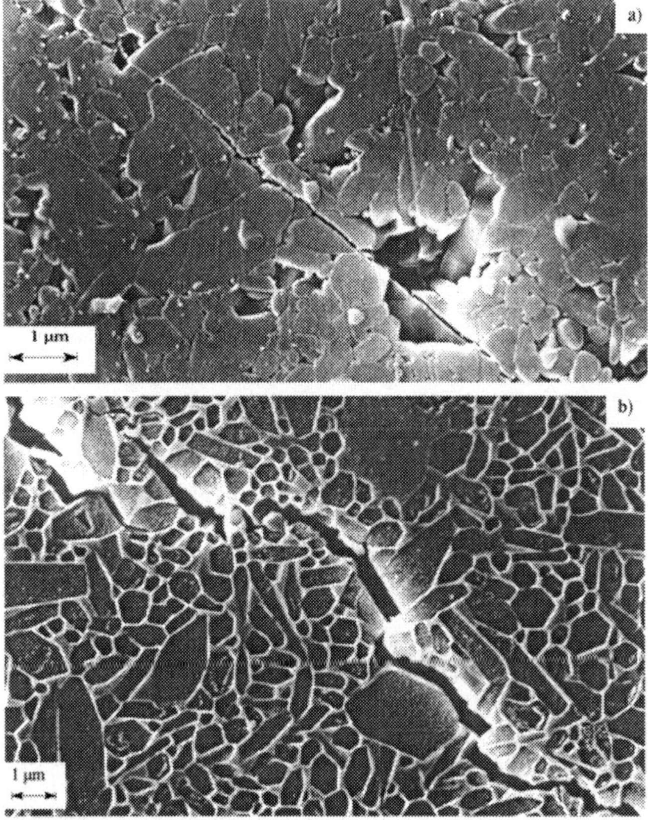

Fig. 29. Crack path in a Si_3N_4 ceramic **a** transgranular with low and **b** intergranular with high fracture toughness

consequence the fracture toughness is high. In contrast, materials with grain boundary phase under compression (e.g., HIPSN without sintering additives) have low fracture toughness due to a high percentage of transgranular fracture (Fig. 29a) [400].

The residual stresses can be influenced during the sintering cycle, either by changes of composition of the grain boundary phase (e.g., by evaporation of SiO_2), or by crystallisation of the glassy phase or by partial relaxation of stresses (e.g., by slow cooling) [357]. These changes are of minor influence and are usually outweighed by grain size, shape and composition, which have a more pronounced influence on stress state and fracture toughness.

The composition of the grains also influences the fracture toughness because of the special chemical interactions that occur between the grains and the grain boundary phase. The formation of Al and O rich βss layers on Si_3N_4 grains, which is especially pronounced in βss materials, results in an increase of transgranular fracture and a decrease in the fracture toughness [33, 401].

αss materials with relatively coarse, elongated grains show a high percentage of intergranular fracture and exhibit a high fracture toughness [332, 333, 337, 342].

The **subcritical crack growth** in Si_3N_4 ceramics takes place only at stresses very close to the fracture stresses, it is low in comparison to other ceramic materials. The growth rate exponent is in the range of n = 30–300 for static loads and n = 20–120 for cyclic loading [293, 396, 402–404]. The growth rates under static and cyclic loading accelerates in water and independent of pH between 2 and 14, similar to that of glasses [404]. Oxide nitride glasses are more resistant to subcritical crack growth than oxide glasses [405]. This, together with toughening mechanisms, may be responsible for the lower subcritical crack growth rate of Si_3N_4 ceramics under static load in comparison to oxide ceramics. Until now, no clear correlation could be made between the microstructure and composition of the material, on the one hand, and the subcritical crack growth on the other hand.

Cyclic fatigue of Si_3N_4 ceramics is more pronounced than static fatigue, i.e., in cases where under static loading no crack growth occurs, pronounced crack growth can take place under cyclic loading. This is connected with the degradation of bridging grains, i.e., the toughening mechanisms can only work partially [403, 404]. The most intensive damage takes place under cyclic loading conditions with alternating compressive and tensile stresses, as a result of the destruction of the bridging large grains. Under cyclic loading, ceramics with smaller grain size have the higher crack growth exponent, i.e., have a lower degradation of strength during loading and a higher life time at a given strength level [402].

The **hardness** of Si_3N_4 ceramics depends on the phase composition because of the different values of the different phases (Sect. 2). Compositions with a high amount of α, which is not transformed during densification, have hardnesses as high as αss (up to 20 GPa) [17, 406]. The macrohardness of different β-Si_3N_4 ceramics increases with decreasing grain boundary phase and grain size [392]; the dependence of hardness on the microstructure is opposite to that of the fracture toughness. Usually the values for β-Si_3N_4 ceramics (HV10) are in the range of 12 GPa (coarse grains, high additive content) to 16 GPa (fine grains, low additive content). The hardness of two phase αss/βss materials changes linearly with the phase ratio [407]. The dependence of the microhardness on the grain size is more complex and depends on the ratio of indentation to grain size [408]. The hardness at elevated temperatures depends additionally on the softening of the glassy grain boundary phase. Materials with MgO/Al_2O_3 as sintering additives show a faster degradation of the hardness than materials with more refractory grain boundary phases.

Si_3N_4 exhibits a high **thermal shock resistance** due to the combination of low thermal expansion exponents, high strength, medium elastic constants, and reasonable thermal conductivity [360, 409], e.g., Si_3N_4 ceramics can withstand quenching in cold water from up to 800–1000 °C whereas Al_2O_3 or ZrO_2 ceramics withstand such a procedure only up to 200–400 °C [360, 409]. Especially *in-situ* reinforced materials have high thermal shock resistance [383].

7.2.2
High Temperature Properties

As a consequence of the strong covalent bonding the properties of the Si_3N_4 grains do not change up to temperatures of 1600 °C but the grain boundary phase already begins to soften at lower temperatures. Depending on amount and composition of the grain boundary phase various processes (diffusion, creep, slow crack growth, oxidation, corrosion) may occur at elevated temperatures (Fig. 28) with the consequence that a new defect population is generated which determines the failure behaviour and limits the lifetime. The extent to which these processes occur is mainly influenced by the softening point and viscosity of the amorphous grain boundary phase [382, 410].

Improvements in all processing steps, and quality of powders allow the increased use of refractive additives, e.g., rare earths which result in remarkable increases of creep resistance, strength and life time of Si_3N_4 ceramics (Figs. 30 and 31) [411–415, 420].

Creep curves of Si_3N_4 at high temperatures generally consist of three regimes: transient, steady-state, and accelerated creep, similar to metals. The creep rate under tensile stresses is some orders of magnitude higher than under compression [412, 416].

Different **creep mechanisms** are discussed in references [383, 412, 413, 416–421]. Diffusional creep is unlikely to be the rate controlling process in Si_3N_4 ceramics with considerable amounts of glassy grain boundary phase. Also dislocation motion contributes only little to creep below 1700 °C. The creep mechanism in Si_3N_4 ceramics is strongly correlated with the grain boundary phase. The dominating processes are material transfer by solution-reprecipitation through the viscous phase and rearrangement by viscous flow [417, 421, 422], formation of cavities [418, 421], or cracks accompanied by

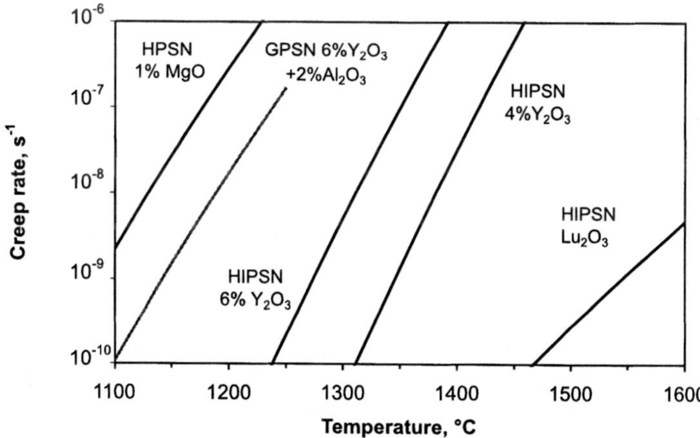

Fig. 30. Creep behaviour of Si_3N_4 ceramics with different sintering additives under 150 MPa static load [413]

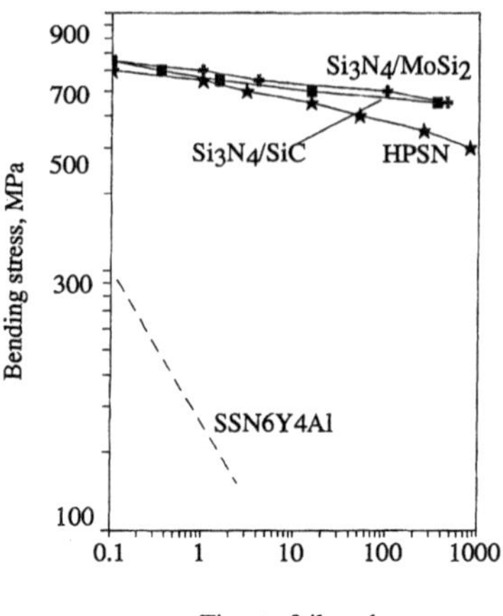

Fig. 31. Time-to-failure behaviour of Si_3N_4 ceramics under bending load at 1400 °C. $Si_3N_4/MoSi_2$; Si_3N_4/SiC composites with 8 wt% Y_2O_3, HPSN Si_3N_4 ceramic with 8 wt% Y_2O_3, SSN6Y4Al shows the typical behaviour of Si_3N_4 ceramic with Al_2O_3-containing sintering additives

grain boundary sliding [417, 420], and redistribution of the secondary phase resulting in cavitated multigrain junctions [412, 416]. The rearrangement by viscous flow leads to a logarithmic dependence of the creep rate on stress and explains the high stress exponents for Si_3N_4 ceramics ($\dot{\varepsilon} \sim \sigma^n$; $\dot{\varepsilon}$ creep rate; σ applied load, n stress exponent) [412, 413, 416]. Thus, creep behaviour of Si_3N_4 ceramics is largely dependent on characteristics and behaviour of the glassy phase. Some disadvantages can be avoided by partial crystallisation of the amorphous phase; this is especially beneficial to creep and slow crack growth behaviour [413].

In HIPSN with no sintering aids the only liquid phase during sintering is the silica adhered even to high-purity Si_3N_4 powders. As the consequence of the small amount and the high softening point of the grain boundary phase, the materials exhibit excellent creep behaviour at temperatures up to 1500 °C [400, 423]. A more refractory grain boundary could only be obtained by removing the silica from grain boundaries and triple junctions. But this is not possible and each addition of only small amounts of impurities or sintering additives will change the chemistry of the grain boundary phase and weaken the material [411, 424]. Besides the costly fabrication, the main disadvantage of these materials is their relatively low strength (500 MPa) and fracture toughness (3–4 MPa m$^{1/2}$) at RT.

Glass-forming and stabilising sintering additives which result in a silicate phase with a low softening point and low viscosity such as MgO, Al_2O_3 or AlN are categorically unsuited for Si_3N_4 ceramics for applications above 1200 °C. Refractory silicates forming intergranular phases with a high crystalline content were obtained by yttria or other rare earth oxides as sintering additives [413, 414, 423, 425]. Superior creep behaviour is found with Lu_2O_3 as sintering additive exhibiting $Lu_2Si_2O_7$ and $Lu_4Si_2O_7N_2$ grain boundary phases [413, 414]. It is supposed that the main creep mechanisms are solution-reprecipitation steps accompanied by significant suppression of cavitation [413]. It is known, that the bond strength of the rare earths increases with decreasing ionic radii, i.e., in the order La...Y, Yb, Lu. However, the change of the glass properties of oxide nitride glasses with changing rare earth additive can not alone explain the drop in creep rate going from Yb_2O_3 to Lu_2O_3 [413, 414]. There are some indications, that the thickness of the grain boundary phase is reduced and that the superior properties are connected with a more complete crystallisation of the grain boundary phases [413, 414]. Fine grained ceramics with grain boundary phases of low viscosity have high creep rates (e.g., 10^{-4} s^{-1} at 60 MPa and 1550 °C) and show superplasticity (deformation without remarkable mictrostructural deterioration) which can be used in shaping [426].

At high temperatures exist two regions in the fracture mechanism map of Fig. 28: slow crack growth failure and creep fracture [413, 414]. The former occurs when a crack grows subcritically from a pre-existing flaw and reaches the critical size. This is predominant in the high-stress, short-life time region and varies with the kind of additives [427, 428]. The creep fracture is due to the formation of a macrocrack of the critical size by cavity nucleation and coalescence, which prevails in the low-stress region and long times. For materials with rare earth additives the change between crack growth failure and creep fracture occurs at tensile stresses of $>200 \sim 300$ MPa at 1400 °C.

For long-term high temperature application, oxidation damage has a significant influence on the lifetime. The migration of sintering additives toward the outer surface, the pore formation in the bulk and the pit formation due to local enhanced oxidation leads to an environmentally caused degradation of the properties. High long term stability in oxidising atmospheres was found for HIPed materials without sintering additives and materials with Lu_2O_3 as sintering additive. A high long-term stability was found also for Si_3N_4/SiC and $Si_3N_4/MoSi_2$ composites having a different oxidation mechanism compared to monolithic Si_3N_4 ceramics (Fig. 31). This is caused by a reduced redistribution of the liquid phase and pore formation in the bulk [410, 423, 429] (Sect. 9.2).

Two phase $\alpha ss/\beta ss$ materials are of interest with respect to their HT behaviour [334, 430]. Due to the ability of the αss grains to incorporate cations from the sintering aids, it is possible to modify the grain boundary. With a large amount of αss, a skeleton of strong grain boundaries between the αss grains is formed due to incorporation of cations in the αss structure, with the consequence of an improved creep resistance. The oxidation resistance cannot be improved by the increase in the αss content; therefore these materials can only be used up to 1300–1350 °C for long-term applications [334].

In principle, the amount, composition and degree of crystallisation of the grain boundary phase are key factors which must be considered for successful development of Si_3N_4 ceramics for applications at elevated temperatures. It is possible to provide Si_3N_4 ceramics for long-term applications up to temperatures of 1500 °C.

7.3
Chemical Properties

Like all other non-oxide ceramics Si_3N_4 is metastable in air or combustion gases, both at room and at elevated temperatures. Detailed understanding of oxidation and corrosion mechanisms and the influence of the surrounding atmosphere on the lifetime are necessary before Si_3N_4 ceramics can be applied under oxidising or corrosive conditions [431–437].

7.3.1
Oxidation

One can distinguish between active and passive oxidation. During **active oxidation** the oxidation products are immediately removed from the surface, causing weight loss, whereas during **passive oxidation** a weight gain takes place, because oxide layers formed.

Normally active oxidation occurs at low oxygen pressure and high temperatures. The transition from active to passive oxidation depends on temperatures and pressures, which are predictable by coupling local equilibria with transport processes (Fig. 32) [431, 436, 438, 439]. With increasing nitrogen pressure the temperature increases and the oxygen partial pressure decreases [438]. In addition, the transition is influenced by the activity of SiO_2, i.e., for glassy surface layers with higher amounts of additional compounds

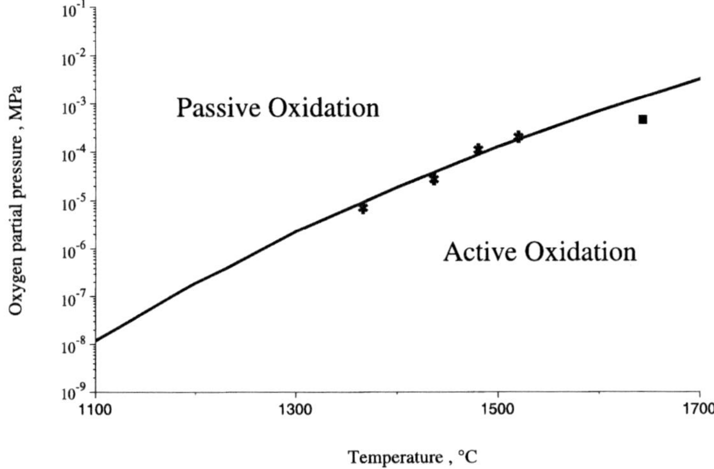

Fig. 32. Dependence of the active to passive transition of Si_3N_4 ceramics on the temperature and oxygen partial pressure [438]

such as Y_2O_3 or Al_2O_3 it takes place at higher temperatures and/or lower pressures [438].

For perfect surface layers the passive to active transition occurs at much lower oxygen pressure than the active to passive transition [431, 439]. In real systems, however, the transition temperatures are identical because of imperfections (cracks, bubbles, etc.) in the layer [438, 440].

In air the active/passive transition takes place at about 1800 °C (Fig. 32). In high temperature oxidation experiments the begin of active oxidation in air was observed between 1600 and 1650 °C, caused by destruction of the layer by bubble formation and evaporation of gaseous reaction products such as $Si(OH)_4$ or $SiO(OH)_2$. The active to passive transition and the accompanying weight loss is altered by gas stream velocity, gas viscosity, laminar or turbulent flow regime of the gas. Beside the oxygen pressure, H_2O, CO or H_2/H_2S in the gas can initiate active oxidation [431, 434, 436]. Active oxidation takes also place in high velocities of water vapour and gas flow below the values predicted in classical theory (Fig. 32) [431, 436, 441–443]. In this case the SiO_2 surface layer reacts with water to gaseous $Si(OH)_4$ or $SiO(OH)_2$ [441]. This is a most serious problem for application of Si_3N_4 ceramics in high temperatures gas turbines. Intensive research on special coatings or surface layers for environmental barrier coatings has not been very successful yet [442, 443].

The different reactivity of phases in multiphase systems results in selective attack and pit formation, reducing strength or causing an enrichment of sintering additives such as Y_2O_3 or Al_2O_3 at the surface [444].

The **passive oxidation** of Si_3N_4 ceramics is mainly influenced by the protective layer formed at the surface by the oxidation process and the ability of this layer to prevent oxygen from diffusing into the material. A surface layer of pure SiO_2 causes a very low rate of oxygen diffusion into the material [445, 446]. However, diffusion of sintering additives or impurities into the oxide layer lowers its viscosity and increases the oxygen diffusion, i.e., it decreases the protection and enhances oxidation [432, 436, 445, 447–449]. Therefore CVD-Si_3N_4 and HIP SN (no additives) have the highest oxidation resistance of all Si_3N_4 ceramics because of their perfect protective layer (Fig. 33). For these materials a diffusion/interface controlled reaction was found [450, 451]. At low oxidation temperature (<1400 °C) in CVD-Si_3N_4 a thin additional Si_2N_2O sublayer with lower diffusion rate has been observed. Presence or absence of an additional Si_2N_2O sublayer is still under discussion and depends on the condition and purity of the material [436, 450, 452, 453]. An amorphous oxide nitride layer with changing oxygen/nitrogen ratio may also occur [453, 454]. No Si_2N_2O sublayers have been observed in additive containing materials, but Si_2N_2O forms in the bulk [445, 449].

In Si_3N_4 ceramics containing additives, the protective layers are complex and composed mostly of the amorphous glassy phase containing some of the sintering additives and cristobalite and silicate phases (Fig. 34) [432, 435, 445, 447, 449, 455]. The temperature of the SiO_2 rich eutectic of the oxide nitride systems including the additives is the limit for long-term applications. Above the eutectic temperature higher amounts of liquid are formed, causing accelerated degradation. For example, the temperature of dramatic oxidation

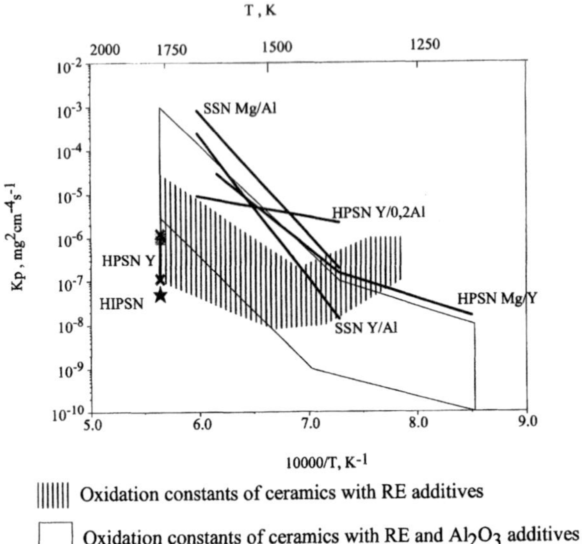

Fig. 33. Parabolic rate constant K_p of different Si_3N_4 ceramics [455], (SSN Mg/Al [465]; SSN-Y/Al [466]; HPSN Y/0.2 Al [466]; HPSN Mg/Y [467]; HPSN Y and HIPSN (no additives) [445]

increase is 1350 °C for materials with Y_2O_3/Al_2O_3 additives while the materials only with Y_2O_3 are stable up to 1500 to 1550 °C [15, 445].

Whereas CVD Si_3N_4 and additive-free HIPSN oxidise at the outer surface, only ceramics containing sinter additives are susceptible to internal oxidation, because of their lowered protection by the surface layer and higher grain boundary diffusion of the oxygen (Figs. 34, 35).

The formation of SiO_2 by the oxidation causes fluxes of cations from the bulk to the surface. In the near surface area an enrichment of additives is found, leading to a damage also of the bulk. Grain boundary diffusion of oxygen into the bulk and the oxidation of the Si_3N_4 grains below the outer surface results in an SiO_2 enrichment, causing fluxes of additives from bulk to surface and an enrichment of additives in the near surface region thus leading to bulk damage (pore formation) and a strong degradation of strength (Figs. 34, 35) [445, 448, 455–458]. This process is especially pronounced at high temperatures, when the protection by the surface layer is reduced and the grain boundary diffusion is faster than the interfacial oxidation [448].

For MgO-containing materials the Mg diffusion is the rate controlling step [459]. The influence of MgO and Y_2O_3 on the oxidation rate depends on the changing amount of cristobalite formed on the interface between the oxide skin and the bulk. The influence of rare earth elements on the oxidation kinetics is much smaller than that of MgO [447].

In composites of Si_3N_4 with $MoSi_2$ and SiC, a moderate damage of the bulk due to oxidation was observed, leading to an increase in life time at high temperature [429, 445, 460]. The reason is the rapid formation of Si_2N_2O near

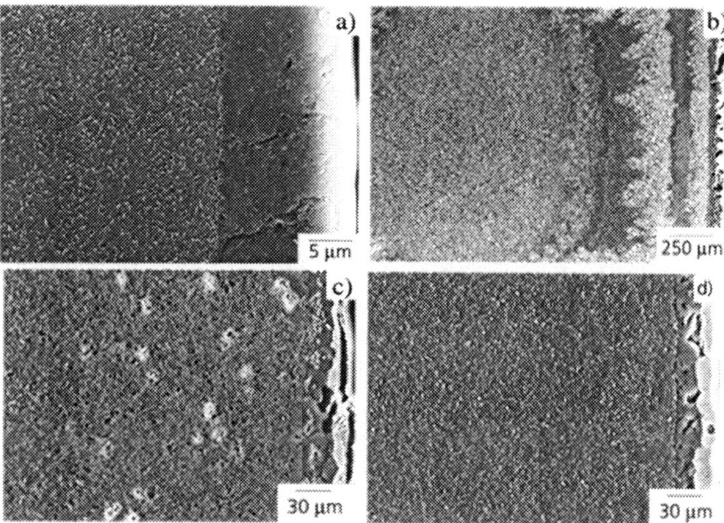

Fig. 34a–d. Oxide layers on Si$_3$N$_4$ ceramics oxidised at 1500 °C. **a** HIP-SN (no additives; 2500 h), **b** SSN (Y$_2$O$_3$/Al$_2$O$_3$ additives; 1000 h), **c** SSN (Y$_2$O$_3$ additive; 5000 h), **d** Si$_3$N$_4$/MoSi$_2$ composite (Y$_2$O$_3$ additive; 5000 h)

the surface, reducing the enrichment of SiO$_2$, which is the driving force for the diffusion of additives from the bulk toward the surface. Additionally, the diffusion of oxygen into the material is reduced.

The oxidation stability of Si$_3$N$_4$ ceramics with different sintering additives at temperatures above 1200 °C increases in the order: MgO; MgO/Al$_2$O$_3$; MgO/R$_2$O$_3$ < R$_2$O$_3$/Al$_2$O$_3$ ≪ R$_2$O$_3$ (Fig. 33). The oxidation rates of ceramics with different additives of rare earth oxides are reduced with decreasing ionic radii, i.e., the oxidation rate decreases in the order La, Ce > Y > Yb > Lu > Sc [413, 461]. Additionally, the oxidation rate depends on the amount of additives and on the additives to SiO$_2$ ratio. So the weight gain during oxidation of Y$_2$O$_3$ containing materials can change by a factor of two, depending on the SiO$_2$/Y$_2$O$_3$ ratio (Fig. 36). Materials with Lu$_2$Si$_2$O$_7$ and Lu$_4$Si$_2$O$_2$N$_7$ as grain boundary phase show a very high oxidation and creep resistance, which cannot be explained by the change of the radius of the rare earth ions in comparison to Y$_2$O$_3$, Yb$_2$O$_3$-containing ceramics. The reason for this high oxidation resistance appears to be a nearly complete crystallisation of the grain boundary phase [413]. The oxidation (parabolic oxidation constants) of ceramics with MgO increases non-linearly with rising MgO/SiO$_2$ ratio at constant amount of additives.

The strong dependence of the oxidation on composition of the ceramic as well as the amount and state of the grain boundary phase (composition and crystallisation ability) is the reason for the scatter of the oxidation constants shown in Fig. 33, i.e., small amounts of impurities may change the oxidation rate by several orders of magnitude [455, 459]. Especially cation impurities

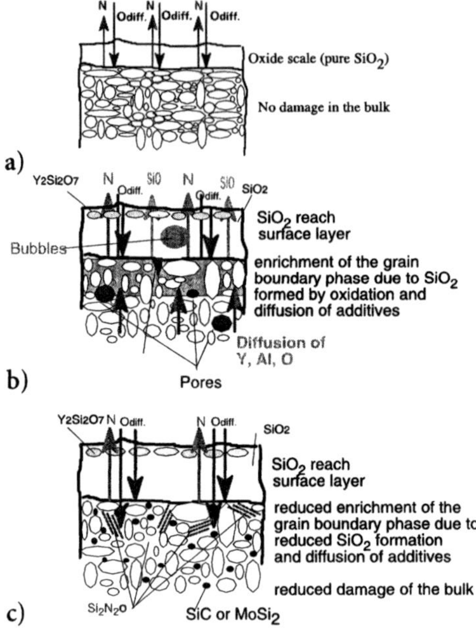

Fig. 35a–c. Schematic representation of the processes during oxidation. **a** HIP-SN no additives, **b** SSN containing Y_2O_3/Al_2O_3 additives, **c** $Si_3N_4/MoSi_2$ composite with Y_2O_3 additive

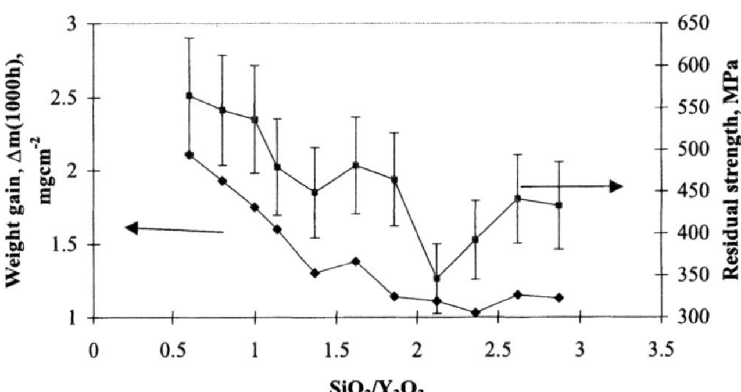

Fig. 36. Dependence of weight gain during oxidation (1500 °C 1000 h) and residual strength after oxidation of HPSN with different SiO_2/Y_2O_3 ratios [446]

such as Na^+, K^+ inhibit the crystallisation and lower the viscosity of the oxide film and reduce the oxidation resistance dramatically [459].

The oxidation rates do not correlate with the strength degradation in all cases. For materials with different SiO_2/Y_2O_3 ratios, the oxidation rate increases with increasing Y_2O_3 content and the residual strength also increases after oxidation (Fig. 36). This tendency is more pronounced for composite materials with $MoSi_2$ or SiC, exhibiting nearly the same oxidation rate as the monolithic materials but a twice the strength after oxidation at 1500 °C. The reason is the different segregation of the grain boundary phase and a reduced pitting tendency (Sect. 9.2) [429, 445, 460].

Different rate laws were proposed for the oxidation mechanisms. The most common is the parabolic law (diffusion controlled process) [436]. For the starting period of the oxidation more complex laws are proposed, including linear, logarithmic or arctan functions of time [436, 462]. The simultaneous crystallisation of the oxide layer, causing a reduction of the diffusion coefficients in the oxide scale, leads to a logarithmic law [436, 455]. Additionally cracks, bubbles and other defects in the surface layer will influence the kinetics and cause deviations from the parabolic law [436, 455, 462]. Ceramics with nitrogen-rich grain boundary phases exhibit accelerated oxidation in the range of 900–1100 °C – the so-called **catastrophic oxidation.** This is caused by the absence of a dense oxide layer [463]. The stresses caused by the volume change during the oxidation of the grain boundary phase lead to cracks and new surfaces undergoing oxidation and finally to a fast destruction. This process increases with increasing additive content, but can be prevented by a short heat treatment between 1200 and 1400 °C to develop a protective SiO_2 surface layer.

The oxidation behaviour of Si_3N_4 ceramics strongly depends on impurities in the gas atmosphere. Impurities like alkaline or alkaline earth metals, SO_2, and vanadium drastically decrease oxidation [431, 433, 434]. The main influence of the different impurities is caused by a change of the viscosity or the destruction of the oxide scale, accelerating the diffusion of oxygen or water vapour into the ceramic and increasing the corrosion. Of coarse, the effect strongly depends on temperature and gas composition.

The corrosion by molten salts was intensively investigated in connection with impurities of combustion gases [431, 433, 434, 436, 464]. The corrosion effect of NaCl in combustion environments is less pronounced, when the sulphur concentration in the fuel is higher [431]. The reason for this behaviour is that at high sulphur concentration Na_2SO_4 is stable and thus the Na_2O activity and the formation of sodium silicates is reduced [431]. These processes are analysed in [431, 433, 436].

7.3.2
Interaction with Metals

Information on the interaction of Si_3N_4 ceramics with metals is important for understanding of the behaviour of metallic impurities during sintering, for the joining to metals [468, 469], for application of Si_3N_4 materials in metallurgy, and as cutting tools [470, 471]. The interactions strongly depend on the nitrogen/oxygen pressure in the atmosphere and on temperature.

In Table 13 the interactions of Si_3N_4 ceramics with common metals are summarised (see also [18, 472, 473]). For application in metallurgy not only the interaction with the metal but also the interaction of oxide slags on the surface of the metals has to be taken into account. At higher temperatures most metal oxides react with the grain boundary phase. For example, V_2O_5,

Table 13. Interaction of Si_3N_4 ceramics with metals

Metal		Chemical interaction	Wetting angle	Remarks
Si		Some solubility of Si_3N_4 in Si: $2\text{-}4 \cdot 10^{-6}$ at % N in Si(l) [474, 475]	Depend on the oxide scale on the surface; at low oxygen pressure wetting angle 45–50 ° [476, 477]; at decomposition conditions of Si_3N_4 wetting was observed [478]	
Alkali metals	Li	Li react at >500 °C forming Li_2SiN_2, Si solves in Li [479]; can cause cracking of the ceramic;		Oxides react very strong with the grain boundary phase; Si_3N_4 films can act as diffusion barriers for alkaline metals in semiconductor devices [481]
	Na	Pure Na does not react up to 1700 K; impurities have an aggressive influence [480];	Na(l) $\sim$20 ppm O_2 at 400 °C–500 °C $\sim$85 °; at lower temperature it is higher	
	K	Pure K does not react [480]		
Earth alkaline metals	Be	No reliable data exist		Reaction of the oxides with the glassy grain boundary phase
	Mg	Reacts only slightly at ≥750 °C; at ≥1000 °C under 100 torr Ar $MgSiN_2$ formation is observed [482]	No wetting of pure Mg at 850–950 °C in vacuum or air 90–150° [482, 484]	
	Ca	Ca SiN_2 can be formed at 1300 °C [483]	Wetting, 10° at melting point [484]	
3a-group metals	Al	Stable with molten Al [474, 485, 486]; different protective layers at the interface can exist of 15-R polytypes [487, 488]; amorphous Al_2O_3; at high temperature AlN formation [489]; Al 7 wt% Mg infiltrate and decompose RBSN [490];dense materials are stable [490];Al/Ca-alloys infiltrate and react with RBSN [474]; no reaction [474, 491]	No wetting near the melting point due to the existence of an oxide layer on the surface; at 900 °C 65° [468] at 1000 °C 126°[484]	Extensively used in aluminium industry

4a-group metals	Ga,	No reaction [474, 491]	No wetting [468, 492]	Used in the GaAs crystal growth technique
	In	No reaction [474, 491]	No wetting [468, 492]	
	Ge,	No reaction	No wetting [473, 496]	
	Sn	Reaction [491]		
	Pb	No reaction in vacuum or protective atmosphere up to 800 °C [493]	No wetting in vacuum or protective atmosphere [491] Pb: 100–150° at 450 °C	Strong reaction of Si_3N_4 with PbO, PbO_2 [493]
3b-group metals	Sc	Depending on nitrogen pressure at 1000 °C ScN, $ScSi_{2-x}$ or Sc_5Si_3 are stable [494, 495]		
	La	Si_3N_4 is stable with molten La; at nitrogen pressure lower than 10^{-4} atm $LaSi_2$ is in equilibrium with Si_3N_4; at about 1600 °C and 1 atm nitrogen $LaSi_2$ and Si_3N_4 are compatible [472]		
3f-group metals	Lan-thanoides	Ce, Ho react with Si_3N_4 under Ar at 1000 °C to form $RESi_2$; at higher nitrogen pressure the nitrides are stable [495]	Gd wets Si_3N_4 [497]	
4b-group metals	Ti, Zr, Hf	Silicide or nitride formation depending on temperature and nitrogen pressure [468, 469; 498]	Due to reaction good wetting [468, 469, 499]	Ti is used as active element in many brazes [468, 469, 499, 500]
6b-group metals	Cr	CrN or silicides are formed [501]		
	Mo	In equilibrium with $MoSi_2$ or Mo_5Si_3 at 1000 to 2000 °C depending on nitrogen pressure [502]		

Table 13. (*Contd*)

Metal		Chemical interaction	Wetting angle	Remarks
	W	W is in equilibrium with Si_3N_4 at 1800 °C and nitrogen pressure >3 MPa , at lower N_2 pressure WSi_2 or W_5Si_3 is formed		
8b-group metals	Fe	Fe reacts at temperatures >700 °C due to formation of solid solution of Si, N in Fe, at higher temperatures silicide formation [503, 504]		
	Ni, Co	Ni and Co similar behavior as Fe [504, 505]		
	Pt, Pd	$PtSi_x$ formation is found >1100 °C in inert atmospheres		Pd is used in high temperature brazes
9b-group metals	Cu, Ag, Au	Stable in absence of oxygen [474]	Poor wetting [474] Cu: 150° at 1100 °C; Ag : 155° at 985 °C [484]	Are used in brazes as non-reactive components

CuO, and PbO react at temperatures $\geq$600–700 °C quite strongly with the grain boundary phase and accelerate the oxidation and degradation. At temperatures below the transition temperature T_g of the glassy phase this interaction can be neglected because of the low ion diffusion into the grain boundary.

The interaction with Fe is the main reason for the unfavourable wear behaviour of Si_3N_4 cutting tools for machining steel. The temperature at the cutting edge during the turning of cast iron is about 800 °C causing only moderate wear. But Si_3N_4 cutting tools are unsuited for steel cutting due to the higher temperature at the cutting edge (>1000 °C) and the resulting fast wear [470, 471].

7.3.3
Corrosion in Liquids

Si_3N_4 ceramics are promising engineering materials for application under corrosive and wear conditions [18e, 506–511]. Si_3N_4 ceramics are heavily attacked by hot acids, hydrothermal conditions, and bases, as can be seen in Fig. 37. The weight loss of pure CVD Si_3N_4 is much lower compared to Si_3N_4 ceramics. The corrosion behaviour of Si_3N_4 ceramics in liquids is determined mainly by the stability of the grain boundary phase. Therefore the corrosion resistance can be altered with the composition by orders of magnitude (Fig. 37) [506, 510, 511].

The corrosion behaviour can divided into a few main classes (Table 14). In most organic liquids (excluding organic acids) no corrosion was observed.

Si_3N_4 ceramics with Y_2O_3/Al_2O_3 additives degrade strongly above RT in medium concentrated HCl, H_2SO_4 and HNO_3 solutions (Figs. 38 and 39). With increasing temperature the corrosion resistance decreases, whereas with decreasing additive content the corrosion resistance increases.

MgO-containing materials have better corrosion resistance in HCl, H_2SO_4, and HNO_3 solutions than Y_2O_3/Al_2O_3 containing materials [506, 510]. The best corrosion resistance in acids was obtained for HIPSN with no additives. Si_3N_4 ceramics are more stable in concentrated (>5 N) than in diluted acids [18e, 506, 507, 516]. H_2SO_4 and HCl do not attack intergranular films as strongly as

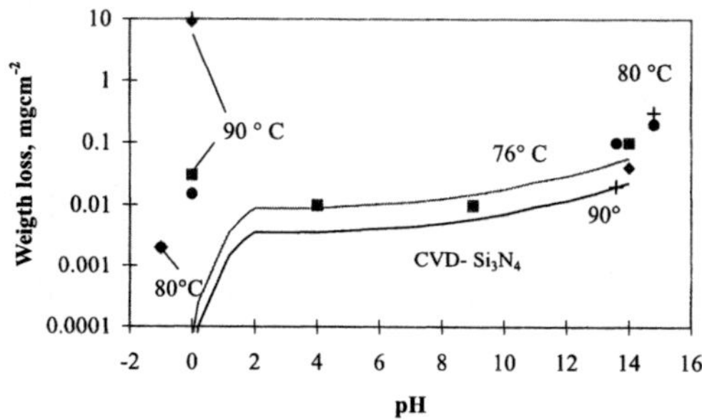

Fig. 37. Weight loss of pure CVD Si_3N_4 [18e] and Si_3N_4 ceramics with different grain boundary phases (measuring points ■, ◆ at 90 °C [525] ; +, ● at 80 °C [506])

Table 14. Classification of corrosion conditions

Conditions	Corrosion	Literature
Organic components (oil, hydrocarbons)	Wear reducing lubricants	
Acids (HCl, H_2SO_4, HNO_3 ...)	Main attack at grain boundaries corrosion resistance can be improved significantly by tailoring the composition	[18e, 506–518]
Media that solves the SiO_2 protective layers intensively	Dissolution of grain boundaries and Si_3N_4 grains	[18e, 513, 515, 518]
(HF, alkaline melts, concentrated alkaline solutions at temperatures >100 to 150 °C); Hydrothermal conditions at ≥250 °C	Intensive corrosion	
Bases at medium temperatures <100 to 150 °C	Main attack of grain boundaries, corrosion resistance can be improved significantly by tailoring the composition	[18e, 507, 510, 513, 519]
Hydrothermal conditions at temperatures ≤200 °C	Main attack at grain boundaries, corrosion resistance can be improved significantly by tailoring the composition	[506, 510, 520–522]

the triple points (Fig. 38). Even after corrosion for 200 h there is no sign, that these boundaries are attacked, whereas the triple points are strongly corroded. Si_3N_4 ceramics with a leached-out grain boundary phase have a low strength of about 400 MPa, similar to RBSN (Sect. 8). The different corrosion behaviour between triple junctions and thin grain boundary films may be caused by deviations in composition; the grain boundary films are richer in SiO_2 [316].

The SiO_2 content of the grain boundary phase is the main parameter governing the corrosion resistance of the Si_3N_4 ceramics. Materials with high SiO_2 concentration in the grain boundaries are more resistant to attack by acids; changes of more than one order of magnitude were measured [511, 517]. Additionally the corrosion is influenced by the fact that corrosion takes place only in small channels between the Si_3N_4 grains, which can cause a change of the rate controlling step (diffusion or reaction controlled, formation of protective layers) [510].

The reduction of strength after short time corrosion in acids (pit formation) cannot be correlated with the weight loss [511], whereas after intensive corrosion a correlation exists between thickness of the corroded layer and strength [507]. Acid corrosion appears to influence subcritical crack growth in Si_3N_4 ceramics [507, 523].

The corrosion resistance of the Si_3N_4 ceramics in **H_3PO_4** solutions differs from that in H_2SO_4 and HNO_3 because a protective phosphate layer is formed [507, 513].

Corrosion in **HF-containing solutions** is much more intensive than in other acids, because of the ability of HF to dissolve the SiO_2 protective layers. HF solutions also dissolve the Si_3N_4 grains.

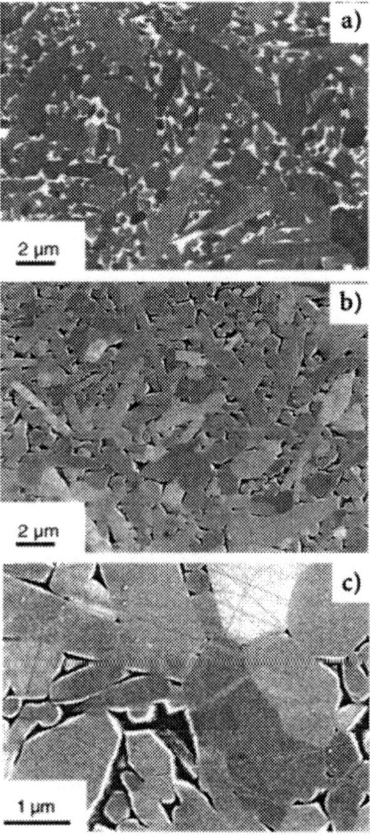

Fig. 38a–c. SEM micrographs Si_3N_4 ceramic with Y_2O_3/Al_2O_3 additives, a before, b and c after corrosion in H_2SO_4 at 90 °C

The corrosion resistance of selected Si_3N_4 ceramics in **NaOH** is given in Fig. 40. The attack by bases is less pronounced than by acids. The extent of corrosion increases with increasing temperature and concentration of the bases [524]. Also, in bases the intergranular films are not attacked as strongly as the triple points. In NaOH solutions often a linear dependence of the weight loss on time is reported. Materials which are less stable in acids are more stable in bases. This can be explained by the stability of the grain boundary phase [510].

Under **hydrothermal conditions** ceramics with Y_2O_3/Al_2O_3 additives are stable (Figs. 41, 42). The HIPed ceramics without additives and the MgO-containing materials are less stable. Under hydrothermal conditions the grain boundary phase of these ceramics is dissolved completely, leading to a removal of Si_3N_4 grains from the surface. This behaviour is different from corrosion in acids, where the grain boundary is not completely dissolved and the corroded layer is quite strong and stable. Under corrosive conditions in

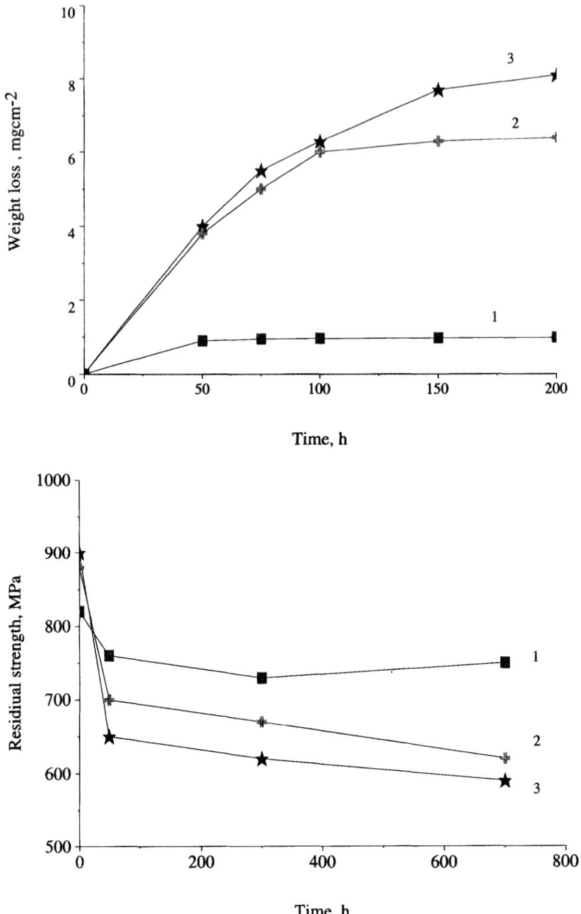

Fig. 39. Weight loss and residual strength of β Si$_3$N$_4$ ceramics with 3.2 (1), 5.0 (2) and 7.3 (3) vol% Y$_2$O$_3$/Al$_2$O$_3$ additives in 1 N HCl at 60 °C as function of time

acids, mainly the network modifiers (e.g., Mg, RE, Ca, alkalines) are dissolved, whereas under hydrothermal corrosive conditions and in strong basic solutions the silica-containing boundary network dissolves. Therefore ceramics with a high SiO$_2$ content in the grain boundary are less stable under hydrothermal conditions than ceramics with a low SiO$_2$ content. Under hydrothermal corrosion at 270 °C a significant dissolution of the Si$_3$N$_4$ grains takes place. In materials with Y$_2$O$_3$/Al$_2$O$_3$ additives, the dissolution rate of the grains is higher than that of the grain boundary [510].

Investigations of the corrosion behaviour of Si$_3$N$_4$ ceramics in acids, bases, and under hydrothermal conditions show that composition and amount of the grain boundary phase govern the corrosion resistance. Materials with high SiO$_2$ content in the grain boundary phase are stable in acids and less stable in bases and under hydrothermal conditions. Materials with a low SiO$_2$ content in

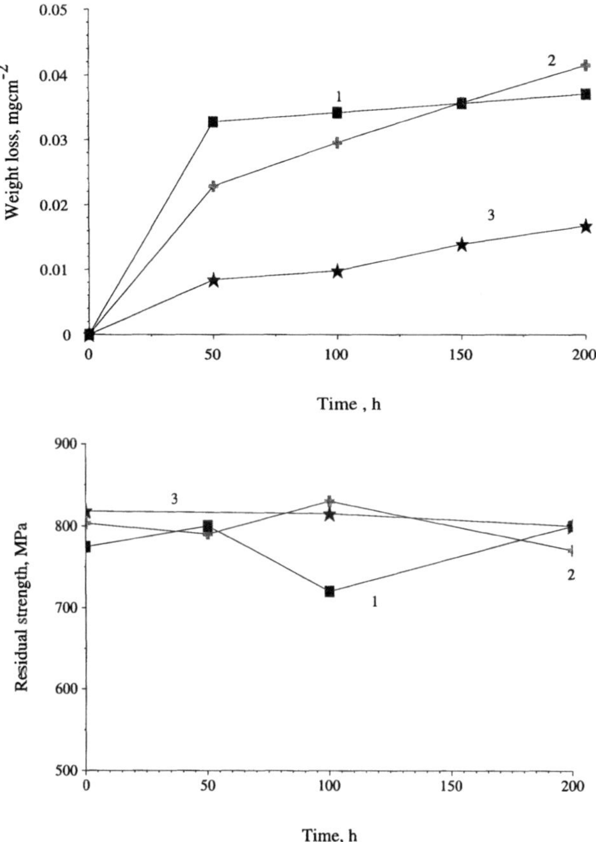

Fig. 40. Weight loss and residual strength of β Si$_3$N$_4$ ceramics with 3.2 (1), 5.0 (2) and 7.3 (3) vol% Y$_2$O$_3$/Al$_2$O$_3$ additives in 1 N NaOH at 60 °C as a function of corrosion time

the grain boundary phase are stable in bases and under hydrothermal conditions but less stable in acids.

7.4
Colours

Si$_3$N$_4$ by itself is colourless due to the big gap between the valence and the conduction band (3.5–5.5 eV) [526]. But normally Si$_3$N$_4$ ceramics are more or less grey, because of the influence of additions, impurities and sintering conditions.

With increasing silicide-forming impurities (d-elements, excluding elements of the third and fourth group of the periodic table of elements) the colour becomes darker. Addition of Ti-containing compounds leads to the formation of TiN. Low concentrations of TiN produce a dark colour, higher concentrations a typical brown to golden colour by the red-brown TiN$_{(1-x)}$C$_x$.

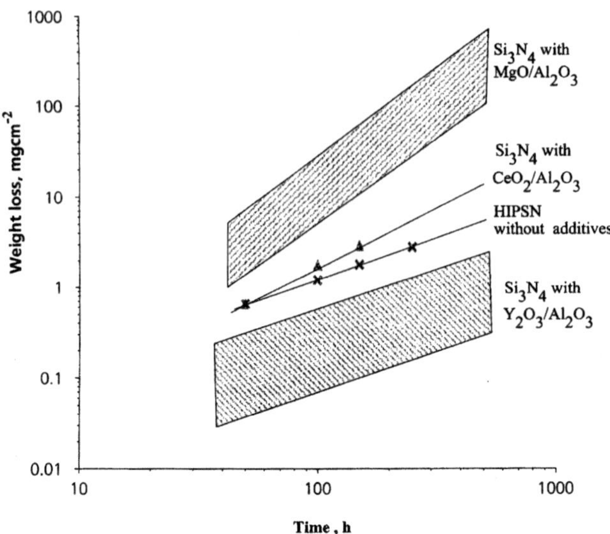

Fig. 41. Weight loss of Si$_3$N$_4$ ceramics with different additives under hydrothermal conditions at 210 °C as function of corrosion time

Under normal sintering conditions Hf or Zr additions form HfO_2 or ZrO_2 (stabilised by REs or MgO) which are bright. Strong reducing conditions generate yellow $ZrN_{(1-x)}C_x$ or read brown $HfN_{(1-x)}C_x$. There are some indications that Cl^- impurities intensify the dark coloration. Impurities of free C lead to a black [232, 272] and SiC additions to a green colour. Additionally, RE ions have colour effects. For example, Yb may colour the material brown or green depending on the oxidation state. The coloration of αss by different REs varies, e.g., violet for Nd [527].

Transparent Si$_3$N$_4$ ceramics were produced by compaction of amorphous Si$_3$N$_4$ under 5 GPa pressure [528]. Translucent αss can be produced by minimisation of the grain boundary phase [335].

The possibilities of colouring described above can be altered by the sintering conditions. Without additional colour centres, Si$_3$N$_4$ ceramics appear between almost white to black.

The intensity of grey colouration corresponds to the free Si precipitates of several nm to several μm which act as colour centres [232, 272, 529]. Impurities of transition metals can be nuclei for precipitation of elemental Si [232, 268, 272], which is most likely formed by the following reaction sequence [232, 271, 272]:

$$3\,SiO_2 + Si_3N_4 \Leftrightarrow 6\,SiO + 2\,N_2 \qquad (17)$$

$$6\,SiO \Leftrightarrow 3\,Si + 3\,SiO_2 \qquad (18)$$

SiO exists not only in the gaseous phase but also dissolved in the oxide nitride liquid. The solubility is quite high, because its structure is very similar

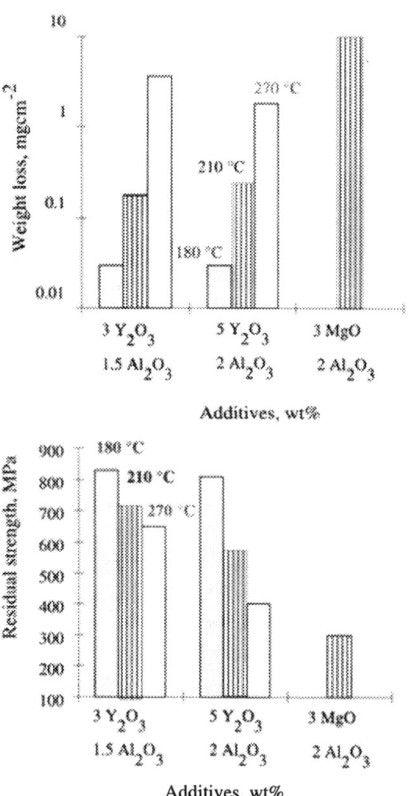

Fig. 42. Weight loss and residual strength of Si_3N_4 ceramics under hydrothermal conditions at different temperatures (corrosion time 200 h)

to that of the oxide nitride melt. Alternatively, this process can be described as a simple formation of oxygen vacancies in the oxide nitride melt or as a simple reduction of silicon from a valence of four to two. Due to the decreasing stability of the divalent Si during cooling from sintering temperatures, the SiO will decompose during cooling according to Eq. (18).

Free silicon is kinetically stable due to the low diffusion rate of nitrogen in the dense ceramic, whereas in ceramics with open porosity, the diffusion path is short enough for the reaction of nitrogen with Si or SiO to Si_3N_4. This agrees with experiments showing that sintered Si_3N_4 ceramics with an open porosity are lighter in colour than those with closed porosity [232, 272].

The grey colour often is not homogeneous and features a bright surface layer followed by a dark layer and a grey coloured bulk material (Fig. 43).

On the basis of experimental data and thermodynamic considerations it can be concluded that the colour of the bulk is less dependent on the weight loss, but strongly depends on the temperature and the nitrogen pressure. With increasing nitrogen pressure during sintering at constant temperature or with decreasing temperature at constant nitrogen pressure the samples become

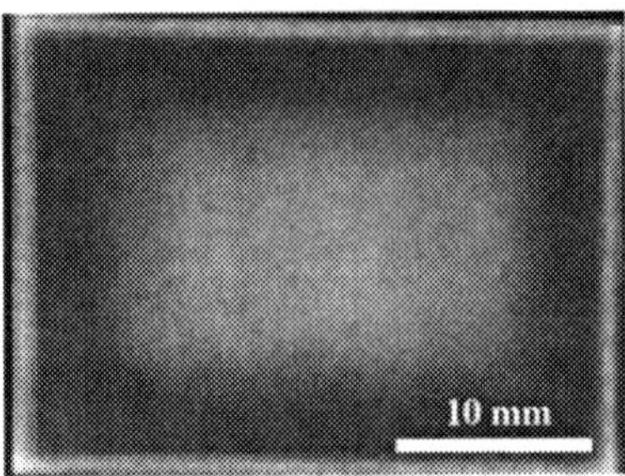

Fig. 43. Optical micrograph of a cross-section through a GSSN rod (Additives : 6 wt% Y_2O_3, 4 wt% Al_2O_3; sintered at 1800 °C; 6 bar N_2; 4 h)

brighter [232, 272]. The dark coloration of the near-surface area depends on the weight loss and the oxygen partial pressure in the sintering atmosphere and is caused by an enhanced reduction of Si^{4+} to Si^{2+} during sintering. The outer bright near surface area results from a decolouration processes during the high pressure step (isothermal sintering), or during cooling from the sintering temperature. The cooling regime will influence significantly the thickness and the colour of the outer layer [232, 272].

Sometimes Si_3N_4 ceramics exhibit the so called "**snow flakes**" with dimensions up to several millimetres (Fig. 44). They appear bright in optical dark field images. Such features are observed in dense HPSN, SSN, GPSN, and SRBSN, independent of the production method. This phenomenon can be explained by local changes in the refractive constants of the grain boundary phase causing a local enhanced light scattering [530]. However, the reason for the "snow flake" formation remains unclear. Several possible mechanisms are discussed: formation of microvoids [531], locally different crystallisation [530, 532] and microcracks caused by volume change during crystallisation of the

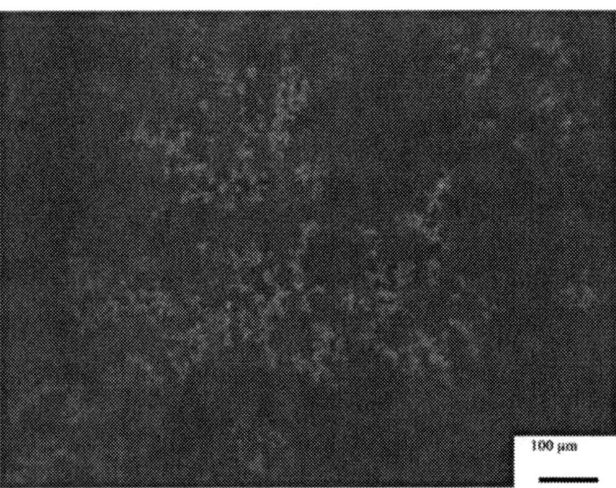

Fig. 44. Dark field image of an silicon nitride material with "snow flakes"

grain boundary phase [533]. For some reason this "snow flake" formation is connected with the crystallisation of the grain boundary phase; all steps leading to a stabilisation of the glass phase, e.g., higher SiO_2, Al_2O_3, MgO contents, faster cooling, reduced interaction with the atmosphere cause a suppression of crystallisation and "snow flakes".

Fe impurities may cause "snow flakes", because of the nucleation and crystallisation of YAG ($Y_3Al_5O_{12}$) [532]. The volume change during crystallisation of the grain boundary phase leads to internal stresses which can be cause micropores, or microcracks, or relax by other mechanisms [534, 535]. Such microcracks have only been detected in ceramics with crystallised β-$Y_2Si_2O_7$ as grain boundary phase [533].

The "snow flakes" can be reduced by heat treatment slightly above the eutectic temperature, even under slightly oxidising conditions [530]. A controlled crystallisation of the grain boundary leads to a more homogenous microstructure. "Snow flakes" are inhomogeneous and cause locally changing microhardness and polishing behaviour [530]. This is especially disadvantageous when a high local mechanical load is applied, as it is the case for ball bearings because these inhomogeneities are potential starting points for pitting damage.

Colouring and "snow flakes" can be controlled by reducing the interaction of Si_3N_4 during sintering, and optimising the temperature, time and pressure regime (Sect. 5.3.3).

8
Reaction Bonded Silicon Nitride (RBSN)

The reaction bonded silicon nitride (RBSN) is the first silicon nitride based material, already produced in the early 50s of the last century [536]. The reaction bonding of Si powder with N_2 results in a material with 12–30% porosity. During nitridation no shrinkage occurs, permitting the production of complex shaped components without expensive finishing. These possibilities and the resulting low cost were the main reasons for an increasing attention and its application as refractory material, especially for thermocouple sheaths and in the metallurgy of aluminium alloys [537]. With reduction in the cost of Si_3N_4-powders and improvements in their densification technology, RBSN ceramics are increasingly being replaced by sintered Si_3N_4 ceramics, which have higher strength, hardness, and long-term stability (Sects. 7 and 10).

By a post sintering process including additional sintering aids RBSN can densified to porosities of about 0–5% [538–540] This material, SRBSN, is applied for different wear parts [540]. A comparison of the production and material properties is given in Table 10.

8.1
Production and Nitridation Process

The production of RBSN starts from Si powder (mean grain size 3–40 μm), which is shaped to parts and then nitrided [541]. The nitridation of the shaped body is the same process as for the synthesis of Si_3N_4 powders by direct nitridation (Sect. 4.1). During nitridation different reactions take place simultaneously (nitridation, sintering of Si, oxidation by oxide impurities of the atmosphere, and evaporation of the adhered SiO_2 layer) influencing the rate of conversion, microstructure and properties [440, 541–546].

Since the reaction is strongly exothermic, local melting of Si might be caused, leading to microstructural defects and reduced strength. Therefore low reaction rates and nitridation times of 24 to 100 h are recommended. Especially in the initial stages a precise control by low heating rates or nitrogen pressures is required.

In the beginning of nitridation a bridging network of Si is built, fixing the overall dimensions of the nitrided part, i.e., no classical sintering process with dimensional changes takes place. Nevertheless the pore volume decreases during nitridation, because of an increase of the solid phase volume of 21.2% by forming of Si_3N_4. The reduction of the pore volume and pore radius causes a reduced transport of nitrogen through pores resulting in a decreased reaction rate with increasing conversion of Si.

Depending on the time/temperature regime, the α/β ratio in the Si_3N_4 compacts varies as a result of varying mechanisms of formation of α- and β-Si_3N_4. α-Si_3N_4 is as fine-grained needles or whiskers in the pores formed by a CVD process of gaseous Si with molecular N_2 [544–546]. Additionally a vapour-liquid-solid mechanism was found, i.e., impurities (e.g., Fe) form a liquid from which the α-Si_3N_4 grows. The whiskers are formed during nitridation of SiO_2 surface layer [440]. All these mechanisms are connected with the evaporation of Si, and all factors that reduce the evaporation also reduce the reaction rate [544]. Therefore the removal of the SiO_2 prior nitridation accelerates the reaction, leading to a coarser microstructure [440, 544]. SiO_2 significantly influences growth rate and density of Si_3N_4 nucleation [542, 544]. Intensive nitridation takes place only when the SiO_2 surface layer is locally removed [440, 541, 542]. Si_3N_4 formed on the surface also reduces the evaporation rate of Si. As a consequence, the nitridation cannot be completed by isothermal nitridation at low temperatures when intensive nucleation of Si_3N_4 occurs.

The formation of β-Si_3N_4 takes place on the surface of solid or liquid Si. The growth of β-Si_3N_4 is epitaxial on the surface of Si [546]. The relatively high diffusion coefficient in the direction of the c-axes of the β crystals promotes growth in this direction. The growth rate is determined by kinetic processes on the interface between Si and Si_3N_4. In liquid Si a fast growth of large β-Si_3N_4 needles is observed. Liquid Si results in an increased β content due to a high nitridation temperature or impurities forming low melting silicides (e.g., Fe forms an eutectic at 1212 °C) [543, 546]. Iron impurities can also accelerate the formation of α-Si_3N_4. Therefore the resulting α/β phase ratio depends not only on the time-temperature regime but also on the amount and distribution of Fe impurities.

Hydrogen or ammonia in the reaction atmosphere increase the α content, the fineness of microstructure, and the strength, they reduce the dependence of the nitridation process on variations in green density and powder properties

[440, 544]. Both gases accelerate the mass and heat transport through the pores and the local removal of the SiO_2 surface layer (it follows from thermodynamic calculations that only in H_2 or NH_3 containing atmospheres can SiO_2 be nitrided [440]). Factors influencing reaction and microstructure are summarised in Table 15. It is obvious that the nitridation is very complex and can modelled only partially at present [547, 548].

8.2
Microstructure and Properties

The development of the microstructure is strongly affected by the perfection of the Si green body. Inhomogeneities in density, pore size and distribution are

Table 15. Influence of processing variables during nitridation on the conversion rate, microstructure and properties of RBSN [541–546, 549, 550]

Processing variable	Product characterisation
Low temperature <1300 °C	High α/β ratio, fine microstructure, normally low conversion rates
Medium temperature 1300–1412 °C	Coarser microstructure with increasing temperature, pose size increases
High temperature >1412 °C	β formation pronounced, coarse structure, low strength and fracture toughness (K_{IC})
Low grain size of Si	Faster nitridation, higher α content, higher strength and K_{IC}
Big grains of Si (>40 μm)	No complete nitridation, reduced strength
Low green density <50% th. density	Faster nitridation (especially of bigger parts), lower strength and K_{IC}
High green density >65–70%	Difficult to achieve full nitridation, strength reduces if inhomogeneous nitridation takes place
No SiO_2 surface layer	Fast nitridation, coarse microstructure, high α content
Thin SiO_2 surface layer (<2 · 10^{-3} g/m^2)	Finer microstructure, high α content, high strength and K_{IC}
Thick SiO_2 surface layer	Difficult to control the nitridation, lower strength and K_{IC}
Fluor addition	Accelerated nitridation, high α content, high strength
Fe impurities (fine dispersed, low content) low temperature	Accelerated nitridation, high α content, high strength
Fe impurities (inhomogeneous)	Defect formation due to silicide formation, reduced strength
Alkaline, alkaline earth oxides, Y_2O_3, Sc_2O_3, RE_2O_3	Accelerated nitridation, high α content, high strength, reduced high temperature strength, subsequent sintering at high temperature is possible
Al_2O_3	Accelerated nitridation, increased β-content
Hydrogen or ammonia in the nitridation atmosphere	More homogenous reaction. High α-content, fine microstructure, high strength and K_{IC}
Flowing N_2 atmosphere	Coarse microstructure, low strength and K_{IC}

unfavourable. The microstructural features normally are open pores (in the range of 12–30%), α/β-Si$_3$N$_4$ grains and residual Si; their size, amount and distribution significantly influence the properties, especially strength. Additionally, the strength may be influenced by defects caused by local melting of the Si or silicides and by large inclusions of non-reacted Si. Whereas the defect size is mainly determined by the green body structure, the fracture toughness is strongly influenced by the nitridation process. A perfect nitridation results in a very fine overall porosity (mean pore channel radius of several 10 nm). This is caused by the growth of fine grained mainly α-Si$_3$N$_4$ into the pores. The strong interconnection between these fine α-Si$_3$N$_4$ grains is responsible for fracture toughness values of about 3 MPa m$^{1/2}$ and strength values of up to 350 MPa. A coarser microstructure leads to a lower fracture toughness and reduced strength.

In inert atmospheres the mechanical properties of RBSN are constant up to 1200–1400 °C because of the absence of a glassy grain boundary phase, which is also the reason for the excellent thermal shock and creep behaviour. The thermal shock resistance, hardness and elastic constants depend on the microstructural parameters but are much lower than for dense Si$_3$N$_4$ ceramics [539].

In air, the mechanical properties are influenced by the oxidation processes [543]. In materials with a fine overall porosity the oxidation at > 1100 °C closes the pores with the help of an SiO$_2$ surface layer. This layer protects the material from further oxidation and heals surface defects. This and the formation of compressive stresses due to the different thermal expansion coefficients between SiO$_2$ and RBSN are the reasons for strength increase after oxidation. Materials with a high amount of macropores (>1 µm) oxidise not only at the surface but also inside the volume due to longer closing times of the surface pores. In consequence these oxidation mechanisms result in more intensive oxidation at low temperatures $\leq$ 1100 °C, due to the slow rate of pore closure and higher internal oxidation.

8.3
Sintered Reaction Bonded Si$_3$N$_4$ (SRBSN)

An remarkable improvement of RBSN is made by including additives in the starting silicon powder and post sintering between 1700–2000 °C after nitridation. During post sintering porosity is reduced and strength increased. The properties of the SRBSN are very similar to SSN or GPSN materials (Table 10) [540, 541, 551]. The additives are the same as for SSN (Sect. 5.1.2). The densification can be attained by pressureless sintering, gas pressure sintering, or sinter-HIPing (Table 10) [538, 552]. The microstructure of the starting RBSN causes some differences compared to sintering of Si$_3$N$_4$ powders. RBSN has a higher green density, but the very strong skeleton of interconnected α needles, as well as the formation of oxide nitride grain boundary phases, noticeably delay the densification due to slow rearrangement mechanisms [552]. SRBSN has the advantage of lower shrinkage and better green machinability. The microstructure formation is very similar to that during sintering of Si$_3$N$_4$ powders.

The strength of the SRBSN is usually between 600–900 MPa and comparable to that of common GPSN. The highest observed values (980 MPa HIP-SRBSN [538]) are much lower than the highest observed values for GPSN materials (1400–1600 MPa). It is an open question whether or not this difference is caused by the lower level of optimisation or by the differences in the sintering behaviour. In general, the properties of SRBSN are very similar to sintered qualities, which can achieved with low cost Si_3N_4 powders. Therefore the former cost advantage of the SRBSN is being lost with fewer applications as a consequence.

9
Composites

To improve, mechanical or electrical properties of Si_3N_4 ceramics different types of composite materials have been developed (Table 16). Among them Si_3N_4/TiN and Si_3N_4/SiC are the most extensively investigated.

Special composites are Si_3N_4 ceramics with additions of β-Si_3N_4 whiskers, or large β-Si_3N_4 seeds are developed to generate β-Si_3N_4 ceramics with bimodal microstructures and improved toughness and strength. They contain two different fractions of Si_3N_4 grains and may considered to be Si_3N_4/Si_3N_4 composites.

9.1
Si₃N₄/TiCN Composites

The interest in composites containing TiN or TiC_xN_{1-x} (x < 1; TiN and TiC form a continuous solid solution with a NaCl structure type) is based on advantageous properties of TiCN such as high hardness, electrical conductivity and chemical compatibility with Si_3N_4. The Si_3N_4/TiN mixtures are stable in a wide range of temperatures and nitrogen pressures [565]; they have good sinterability without decomposition. Details about the solid solubility of Si in TiN of the Si_3N_4/TiN composites without additives differ considerably: up to 2.3 wt% [566] and 10.7 wt% Si [567]. No solubility of Si in TiN was observed in Si_3N_4/TiN composites produced by hotpressing of Ti- and Si-containing organometallic precursors. Although a clear decision between this contradictory data is not possible, a low solubility of Si in TiN is more likely. With respect to the solubility of TiN in β-Si_3N_4 an analogous situation exists. From EDX investigations, and confirmed by a slight change of the lattice parameter, a solubility of 0.6 wt% of TiN in β-Si_3N_4 has been determined [567]. But this is not in agreement with the findings of submicrometer-sized TiN precipitates in composites with only 0.2 wt% TiN prepared by liquid phase densification, indicating a solubility of TiN in β-Si_3N_4 below 0.2 wt% [525].

TiC is thermodynamically not compatible with Si_3N_4 under sintering conditions. TiC reacts with Si_3N_4 forming TiC_xN_{1-x} where the x values are in the range of 0.3 [568, 569] to 0.24 [567]. The x value in TiC_xN_{1-x} depends on nitrogen pressure and temperature. As an additional reaction product, SiC may be formed or the oxide nitride liquid phase can be reduced by carbon

Table 16. Overview of Si_3N_4 ceramic composites

Composite	Improved properties	Remarks	Literature
Si_3N_4/BN	Wear behaviour and machine stability, resistance against metals, high thermal shock resistance. Hot pressed nanocomposite with high strength and improved machinability	BN retard the sintering, materials with higher BN content are difficult to sinter	[553–556] [366]
Si_3N_4/TiN/TiC	High electrical conductivity, wear behaviour	Sect. 9.1	
Si_3N_4/SiC	High mechanical properties at RT and HT	Sect. 9.2	
Si_3N_4/Si_3N_4	High fracture toughness and strength, thermal conductivity	Sect. 9.3	
Si_3N_4/TaN	Electrical conductivity	TaN react under sintering conditions forming silicides and mixed nitrides	[557, 558]
Si_3N_4/ZrO_2/HfO_2	Only moderate increase in toughness due to transformation toughening; reduced hardness and oxidation stability with higher ZrO_2 content.	Under strong reducing conditions $ZrN_{1-x}C_x$ or $HfN_{1-x}C_x$ can be formed leading to instability	[559, 560]
Si_3N_4/TiB_2	Only moderate improvement of K_{IC} or cutting behaviour	Thermodynamically not stable under sintering conditions; decomposition of TiB_2 in TiN and BN or $TiSi_x$	[555, 559]
Si_3N_4/$MoSi_2$	Significant improved oxidation resistance and life time at high temperatures, electrical conductive	Depending on temperature and nitrogen pressure Mo_3Si_5 can be formed during sintering	[410, 561, 562]
Si_3N_4/W	W wire in Si_3N_4 for heating elements	Depending on nitrogen pressure and temperature WSi_2 can be formed leading to a destruction of the wire	[563]
Si_3N_4/other silicides	Less important due to less stability against oxidation or low melting point	Under sintering conditions liquid silicides can concentrate in big defects, Fe silicide can improve the wear behaviour in engine applications	[564]
βss/Al_2O_3/carbon	Increased stability against metals used for refractories	Al_2O_3 particles partially dissolve during sintering, materials usually not dense	[555]

[568, 569]. Therefore the TiC-containing materials are more difficult to sinter and are used to a lesser extent [568, 570].

Most of the composite materials were produced by mixing TiN or TiCN powders with Si_3N_4 [559, 569, 570]. Also additions of nanosized TiN particles to Si_3N_4 was tested successfully [571]. The nitridation of mixtures of Si and Ti or TiN results in an Si_3N_4/TiN particulate composite [572].

To describe the formation of composites with low TiN content a simple exchange reaction can be used:

$$3\,TiO_2 + Si_3N_4 \Leftrightarrow 3\,TiN + 3\,SiO_2 + 1/2\,N_2 \tag{19}$$

The result is a composite with a very fine homogeneous distribution of TiN of nanometer size in the Si_3N_4 matrix. Higher starting amounts of TiO_2 result in Si_2N_2O formation [525, 573]. The composites can also be prepared from metalorganic precursors [574].

The densification of TiN containing composites begins at lower temperatures than the densification of the pure Si_3N_4 matrix. The reason is that the solution of TiO_2, which adheres to the surface of TiN powder particles, reduces the viscosity of the liquid phase. At temperatures >1350 °C TiN precipitates from the liquid according to Eq. (19) [568, 570].

The thermal expansion coefficient of TiN and TiC_xN_{1-x} is higher than that of Si_3N_4. Therefore around the TiCN grains, local tangential compressive and radial tensile stresses are formed. These stresses change the crack path and increase the fracture toughness. Systematic investigations of the dependence of the fracture toughness on TiN grain size and content has shown that with increasing particle size and volume contents up to 30 vol%, the fracture toughness increases slightly without strength increase [568, 570, 575]. This is connected with the intrinsic high fracture toughness level of the Si_3N_4 matrix [559, 570, 575]. Nanosized TiN particles in the composites give no improvement of the properties compared to composites with TiN particles in the μm range [571].

Composites with more than 30 vol% of TiN have a high electrical conductivity, which is very useful for the final shaping by electrical discharge methods or other applications where the electrical conductivity is involved [559, 576, 577]. The specific resistivity depends on the grain size of the TiN particles and changes drastically between 20–30 vol%, expressed by a drop from high values (10^{13} Ωcm^{-1}) to low ones (10^{-1}–10^{-3} Ωcm^{-1}). This drastic drop shifts with decreasing particle size to lower volume contents of TiN [577]. The formation of duplex microstructures, consisting of TiN free areas of up to 200 μm surrounded by TiN, results in materials with high electrical conductivity down to low TiN contents (9.8 vol%) [578].

The oxidation resistance of TiN and TiCN is lower and therefore the composites are less stable at high temperatures than the monolithic materials. When the TiN particles do not form a continuous skeleton, then only surface TiN particles oxidise rapidly. The further oxidation is determined by the stability of the matrix. Composites with a TiN skeleton oxidise more readily than the Si_3N_4 matrix [559, 570, 579].

The composites have an improved wear behaviour under load [580–582]. In comparison to monolithic Si_3N_4 ceramics, the differences at 800 °C were more pronounced than at room temperature. This was attributed to the formation of TiO_x layers, which are known as solid lubricants [580, 581].

Also, materials with small amounts of TiN additions show lower friction in ball bearing applications leading to higher loading capacities of the bearings [525, 583].

Si_3N_4/TiCN composites were tested for cutting tool applications under different conditions with little success [559, 570].

Layered ceramic Si_3N_4 composites on the basis of TiN or TiCN have been prepared by tape casting and hot pressing and showed highly anisotropic electrical and mechanical characteristics. One idea behind this development is the detection of crack formation under loading conditions by electrical conductivity measures [584, 585].

9.2
Si_3N_4/SiC Composites

Already in the early 70s of the last century the first investigations of Si_3N_4/SiC composites were carried out [586]. Unfortunately the strong interaction of SiC with the liquid phase during sintering causes deterioration of the densification behaviour [275, 586–588] and the interest in this material declined. In the late 1980s new investigations in connection with SiC whisker reinforcement began. But the increase in fracture toughness was not substantially greater than that of the *in-situ* reinforced monolithic Si_3N_4 ceramics [589–593]. Additionally, the carcinogenity of the SiC-whiskers was a strong argument for discontinuing the research activities [594]. The incorporation of SiC platelets instead of whiskers yields only a moderate improvement in the properties [595, 596]. With the beginning of the 1990s the interest in Si_3N_4/SiC particulate composites was rekindled in connection with the nanocomposite concept [223, 587, 597, 598]. The improved quality of the powders and advanced technology enabled development of materials with high strength at room and high temperatures [223, 597]. These promising results have intensified the investigation of powders, their preparation and the properties of the resulting composites [213, 587, 598].

The recently developed Si_3N_4/SiC composites are based on

- amorphous or crystalline composite powders [599–603],
- mixtures of Si_3N_4 and SiC powders [587, 604, 605],
- *in-situ* synthesis of SiC during sintering [606, 607],
- polymer derived routes [213, 608–611].

Most of the nanocomposite materials were produced by hot pressing [223, 275, 587, 597–599] to overcome the difficulties during densification. For gas pressure sintering and HIPing, the decomposition reaction of SiC under high nitrogen pressure and the interaction of SiC or residual free carbon with the oxide nitride liquid during sintering must be taken into account [275].

However, gas pressure sintered high strength composites were also developed [275, 604, 612, 613].

The SiC particle distribution is intergranular and intragranular. The ratio of inter- and intragranular particles depends on the SiC grain size and the growth conditions of the Si_3N_4 grains [223, 275, 587, 597, 612, 614, 615]. There are SiC particles with no glassy phase between Si_3N_4 and SiC grains and others which are surrounded by a glassy phase [615, 616]. In materials with an SiC content up to 30 vol%, size and shape of the Si_3N_4 grains are similar to those in monolithic material, and the SiC particle size distribution is broad between some nm up to 1 µm. The SiC particles reduce the exaggerated grain growth by pinning (Zener mechanism), but the overall grain size changes only slightly. The pinning effectivity increases with decreasing grain size and increasing volume content of SiC. In composites with high SiC content (>50 vol%) produced from amorphous Si_3N_4/SiC powders the SiC and Si_3N_4 particles are nanosized [223, 597, 599]; these materials are superplastic [617].

Ceramics with fine SiC and Si_3N_4 grains can also directly produced from amorphous precursors with no sintering additives, however they are not dense after crystallisation and therefore have relatively low strength and fracture toughness [213, 609–611].

The strength of Si_3N_4/SiC ceramics increases compared to the monolithic state especially with Y_2O_3/Al_2O_3 additives [223, 597, 600, 606, 611, 612]. The positive effect on the strength is connected with the pinning mechanism, reducing the accelerated growth of big grains which are strength limiting defects in high strength materials [587, 605, 612]

The wide scatter of the maximum values of fracture toughness [587, 598] or sometimes even a reduced fracture toughness by adding SiC [587, 598, 599, 612] indicate that changes of powder characteristics, processing variables, or of the composition of the grain boundary phase are of more influence than special mechanisms connected with the nanoparticles [587, 604]. The increased residual stresses [604], caused by the different thermal expansion coefficients of the different phases, might not be the reason for the changes in fracture toughness and strength [587, 604] as assumed earlier [606].

The improvements of high temperature strength and creep resistance are mostly connected with shift of composition of the grain boundary phase caused by the interaction of SiC or residual free carbon with SiO_2 during sintering [587, 604, 612, 618]. For instance the high temperature strength of Si_3N_4 ceramics is substantially affected by the crystallisation of the grain boundary phases (Sect. 7.2.2), which is influenced by the SiC addition (shift of the composition of the grain boundary phase) [587, 619]. Materials with constant compositions of the grain boundary phase have very similar high temperature strength [605, 618, 620]. In addition to the change of composition of the grain boundary phase, the SiC particles in the grain boundary have a positive influence on creep. SiC particles without a glassy phase between SiC and Si_3N_4 particles form rigid skeletons, reducing the grain boundary sliding [587, 612, 616]. In HIPed materials with SiC inclusions but without other additives the creep rate decreases only by a factor of 2 [621]. This change seems to be connected with a true strengthening effect by the SiC particles.

Higher differences between the creep rate of the composites and the monolithic materials might be caused by changes in the grain boundary phase, impurities or grain size [429, 460, 622, 623]. Only slight altering of existing creep mechanisms known from pure materials were found [604, 605, 621].

In Si_3N_4 ceramics with SiC additions the residual strength after oxidation increases because of a change in the oxidation mechanism [429, 460, 605, 624]. Just below the outer oxide layer a protective layer of Si_2N_2O is formed which reduces the damage in the bulk (reduced migration of additives toward surface, and pore formation) and the size of pits created by oxidation [429, 460, 605]. Similar behaviour was found in $Si_3N_4/MoSi_2$ composites [429, 620]. This changed oxidation mechanism is also the reason for higher life times, under load, of such materials at high temperatures [460, 605]. No differences in the wear resistance between Si_3N_4 and Si_3N_4/SiC composites was found [625]. Si_3N_4/SiC nanocomposites with SiC contents>25 vol% have reduced electrical resistivity ($<10^7$ Ωcm) [610, 611, 626].

Summarising the data about the Si_3N_4/SiC composites, it can be stated that no significant increase in the mechanical properties at room temperature has been achieved which can not be realised in monolithic Si_3N_4 ceramics. However, at high temperatures substantially improved long term behaviour connected with a change in the oxidation mechanism can be realised.

9.3
Si_3N_4/Si_3N_4 Composites (Seeded Materials)

High fracture toughness of Si_3N_4 ceramics is obtained by activation of toughening mechanisms such as crack-deflection or bridging via interfacial debonding at large elongated grains [33, 310, 377]; but large elongated grains can also act as strength limiting defects (Sect. 7.2). To overcome this discrepancy, ceramics with large elongated grains of narrow size distribution embedded in a fine grained matrix with oriented rod-like grains are desirable. Such ceramics can be realised by large elongated β-Si_3N_4 crystallites (usual length 2–10 µm; thickness 0.5–1.3 µm; aspect ratio 4–5) as seeds [299, 377, 627–635]. Such seeds usually are prepared by sintering of Si_3N_4 powders with Y_2O_3/SiO_2 additives at 1850 °C, subsequent acid treatment to remove the additives and classification by size [629, 633, 635]. The seeds (whisker-like large β nuclei) grow anisotropically consuming the fine grains of the matrix as described in Sect. 6.1. The amount of the seeds should be moderate to avoid impingement of grains. The best results were observed with amounts between 2 and 5 vol% [299, 377]; larger amounts (>10 vol%) cause a fast consumption of the matrix grains by the growing seeds. Exaggerated grain growth results in a coarse grained microstructure with a less pronounced anisotropic shape [299]. With tape casting [377, 629, 632,] or extrusion [631] the seeds can be strongly oriented in the starting materials as a precondition for an optimised anisotropic order. The degree of order of the seeds is the higher, the smaller the amount of seeds [629]. For instance, tape casting of seeded Si_3N_4 powders leads to a high degree of orientation of the seeds (>60%) and hence to anisotropic fracture toughness and strength (5 MPa m$^{1/2}$ parallel to the grain alignment and 12 MPa m$^{1/2}$ normal to it). In tape cast materials a simultaneous increase of strength (from 1100 to 1400 MPa) and fracture toughness from 7.1 MPa m$^{1/2}$ to 12.5 MPa m$^{1/2}$ by addition of 5 vol% seeds was observed [377]. Such ceramics with oriented grains also have high thermal conductivity, up to 160 W(mK)$^{-1}$.

Seeding can be used also in reaction bonded materials (Sect. 8). Increased fracture toughness and strength were observed in reaction bonded and post sintered materials [633].

An increase of the fracture toughness from 2 to 6 MPa m$^{1/2}$ was observed in hot pressed Si_3N_4 ceramics with 50 vol% residual α-Si_3N_4 by adding 5 vol% seeds [635].

10
Applications

During the last 40 years Si_3N_4 ceramics have developed from exotic to commercial materials with increasing application. At present, cutting tools are the most important market; followed by engines components, ball bearings, metal forming and processing devices, and gas turbines. An overview on the applications of Si_3N_4 ceramics is given in Table 17. A rough idea of the market and the share of the different applications can be derived from the powder consumption. One third of the total production of Si_3N_4 powders (300–350 t in 1998) was used for cutting tools, 25% for engine components, 25% for metal processing and wear parts and 2% for ball bearings. The remaining 10% are used in research [636].

Si_3N_4 ceramics cover only 1% of the total market of advanced ceramic materials, i.e., electronic and structural applications, but about 5% of the structural ceramics. They have the highest growth rates among structural ceramics [636–638].

The use of Si_3N_4 ceramic components in engines is now seen more realistically than in the past. The adiabatic full ceramic engine, as it was proposed, has not been realised, but there are ongoing efforts to improve the design of this engine [648, 654]. Also, the production of more than 300,000 turbochargers per year in the beginning 1990s in Japan did not result to series introduction in passenger cars [430, 654]. But the development of components for different engines are still under way. Injector links, check balls, brake pads and fuel pump rollers are used in diesel engines for trucks and have been produced in several thousand pieces per month for years [646, 647]. These applications are not really at high temperatures, but very beneficial due to reduced wear and weight. Also in a newly developed high pressure common-rail injection pump system, valve plates made of Si_3N_4 ceramics are used to reduce wear [402].

The production costs are a main problem in the application of Si_3N_4 ceramics. The production of small series is more expensive compared to existing materials. This is connected on the one hand with the high costs of raw material and on the other hand with the production technologies. In

Table 17. Applications of Si_3N_4 ceramics

Area of Application	Application	State of the market	Literature
Cutting	Cutting tools for turning and milling of cast iron; αss/βss for turning and milling of Ni- based superalloys; Granulation of polymers, cutting of textiles and fibres	Stable market; approx. 50 million \$/year; moderate growth	[471], [637], [639]
Bearings and sealing	Ball bearings especially for high speed high stiffness applications; No or low lubrication (vacuum pumps, dental drill handpieces); Under corrosive conditions (food ndustry; chemistry; metallurgy); high and low temperatures	Fast growing market for hybrid and ceramic bearings 10 times increase in hybrid bearings between 1995–1999. Increasing numbers of balls reduces the price. Development of other rolling bodies is in the beginning	[583], [637], [640], [641]
Metallurgy	Different parts in Al metallurgy (components of pumps working in different metals Al, Pb); locating pins and other parts for welding operations	Growing market, RBSN is more and more replaced by low cost SSN	[636, 637], [640]
	RBSN refractory materials	20 t/year	[14]
	Springs, jigs for brazing of electronic components, jigs for glass sealing,	Small market	[637]
Metal forming	βss Rollers for cold rolling operations	Small market	[430]
	Punches, dies, plugs, mandrels (Fe; Al; Cu)	Small, but fast growing market	[642], [643]
Automotive	Valve plate in common rail systems	Several million pieces/year	[402], [637]
	Glow plugs for diesel engines and additional car heating	Increasing market	
	Fuel injector parts, valves, valve train components, rocker arm pads, tappet disks, cam follower roller, turbocharger rotors; tappet shim precombustion chambers	Used in trucks. Components in test cars and limited series	[645–649]

Aircraft engines	Ceramic turbine nozzles in the APU (Auxiliary Power Unit Engine).	More than 65,000 h and 5500 starts in field tests.	[650, 651]
	Seals runners installed in different business jets such as Falcon, Citation and Learjet. Ceramic oil pump spacer enhance engine could start capability on APUs on Boeing 777 and 737 and all Airbuses.	Growing market: better specification and standardisation needed.	
	Cutter pins, which are used as a safety feature for air turbine starters. Ceramic wear indicator on Airbus APU starter brushes since 1997	More than 30,000 cutter pins are installed	
Space technology	Turbo pump of the space shuttle; radar windows for rockets	Small market	[644]
Electronic industry	Thin films for insulating barriers or masks for etching processes; XY machine tables for VLSI semiconductor manufactures; vacuum jigs	Increasing market	[14]
Chemistry/oil industry	Different chemical machinery parts as valves sealing,	Growing market,	[642]
	Microfilter; high strength, chemical resistant porous materials	No commercial application at present	[652, 653]
	Wear parts (mill liners, milling balls)	Small market	[642]
Gas turbine	300 kW output power turbines	Test runs	[654, 655]
Household	Cooking plates with integrated heater for electrical cooking for kitchens and camping (low voltage systems)	Introduction in the market at present	[656]

recent years, substantial improvements in the production technology have been made and it was shown that mass production can lower the cost substantially [644]. The improvement of properties and production technology result in an increasing application of Si_3N_4 ceramic components in engines. They can be produced with high reliability, as demonstrated in extensive field tests of valves carried out in several thousand 1600 C200-ML cars by Daimler-Chrysler in more than several million miles [644, 649]. The higher production costs of the ceramic valves compared to steel valves and also changes in engine design are reasons that mass production has not been achieved to date. There still exist test cars of low weight and high efficiency loaded with Si_3N_4 ceramic valves [644, 649].

The high reliability of Si_3N_4 ceramics is demonstrated by hybrid bearings in the main engine pump of the space shuttle [644] and different components in the aircraft's auxiliary power units [650, 651].

The application of Si_3N_4 ceramics as components for gas turbines is under development. First successful tests of those components have been carried out with 1350 °C turbine inlet temperatures. The further improvement of the turbine efficiency requires higher turbine inlet temperatures; for this, additional coatings are necessary (Sect. 7.3).

Si_3N_4 thin films and coatings are finding increasing application in electronics as electrical diffusional barrier or as masks for etching processes. Si_3N_4 coatings produced by plasma-enhanced chemical vapour deposition are increasingly used in silicon photovoltaics as passivating antireflection coating [657].

Low cost Si_3N_4 powders and advanced processing technologies facilitates the production of materials with 800 MPa bending strength for cooking plates with integrated heater. These plates are the central component of an effective cooking system allowing automatic control of the cooking process [656].

An interesting application of Si_3N_4 powders is as a foaming agent in the production of glass foams for thermal insulating in uses ranging from microelectronic devices to fire resistant non fibre containing insulation materials for buildings [658]. Low-grade Si_3N_4 powder is used in the steel industry to increase the nitrogen content of the metal; this could also be a possible application for recycled Si_3N_4 ceramics.

The largest quantity of Si_3N_4 by weight is applied in nitride bonded silicon carbide refractories which are produced in amounts of 20,000 t/year [14].

11
Conclusions

Si_3N_4 ceramics represent a whole class of different compositions with many facets and a wide range of properties, and as a consequence they have a high potential for specific applications. They are light, have good mechanical and thermomechanical properties and they are wear and corrosion resistant. Differences between the individual types are founded in grain morphology, and amount and chemistry of the grain boundary phase. Not accidentally the comparison is made to the broad variety of steels which satisfy a multitude of technical requirements. After more than three decades of intensive research good progress has been made in the scientific and practical exploration of these sophisticated materials which characteristically combine all advantages and all problems of advanced ceramics. This is particularly true of all the sintered Si_3N_4 ceramics (SSN, GPSN, HPSN, HIP-SSN, HIP-SN) which are suitable for many applications not attained by the reaction bonded grades (RBSN, SRBSN, HIP-RBSN). Nowadays, Si_3N_4 ceramics can be processed with high reliability and can be adjusted over a broad range of properties for desired applications. However, this state of the art is based mainly on empirical experience. A quantitative understanding of the whole chain of events from the production of powders, their densification up to the microstructure and its connection with properties of the components has yet to be achieved.

Si_3N_4 ceramics are multicomponent systems of higher order and it is obvious that the corresponding phase diagrams are very powerful tools in understanding the relations between processing, microstructural development and final properties. In establishing these interrelations, all chemical constituents and phases must be considered; even small amounts of impurities can produce large effects. So far, all available systems have been investigated only partially, and more than a few experimental results have been misinterpreted. Therefore, a systematic study of constitutional phase diagrams of Si_3N_4 ceramics, including their promising composites, is necessary and this should be combined with thermodynamic calculations of the equilibria and metastable states.

The advances that have achieved so far are mainly based on increasing understanding of processing and microstructure/property relationship. New analytical methods and high resolution transmission electron microscopy have provided new insight into the grain boundary region and offer the possibility to tailor the microstructure for specific applications.

But in the area of specific microstructural engineering more knowledge is desirable concerning the interplay between the microstructural features – Si_3N_4 (ss) grains, secondary phases (in general oxide nitrides) and amorphous phase – during sintering heat treatments and long-term high temperature exposure to avoid improper deviations and defects.

Especially the phenomenon of the amorphous grain boundary phase has not been clarified completely and is a challenge for basic research. Similarly open questions exist concerning the optimisation of fine-grained and nanoscaled materials for applications under extreme conditions. Here, the recently developed fine β-powders may have greater promise than the commonly used α-rich powders. The correlations between microstructure wear, creep, corrosion and long-term behaviour are by far less understood than the correlations between microstructure, strength and toughness. More intensified research on this topic should include the Si_3N_4 composites which may have some advantages. Concerning improvement of fracture toughness αss and αss/βss compositions are the prime candidates for study, their mechanical properties can be substantially improved by tailoring their fibrous microstructure.

The possibility to produce Si_3N_4 ceramics of precise composition and high purity from low molecular weight, inorganic or organoelement precursors is a topic of increasing relevance for small and thin components.

Si_3N_4 ceramics have not only the potential to compete successfully with other engineered materials, but also to initiate new technical concepts which have not caught on for lack of suitable materials. On the other hand, the introduction to the market is slow and Si_3N_4 ceramics still are a niche market. Cost still remains a major barrier. To overcome this handicap a cost reduction is urgent. In this connection, a less-expensive raw material for powder production and lower grade qualities of the starting powders are decisive factors. Furthermore, improving the steps in the production and densification processes may cut the costs. These include intimate mixing of Si_3N_4 powders with additives, avoidance of agglomerates, and homogeneous particle distribution in the green state, pore minimisation during densification and elimination of all defects occurring in the various stages. Lower sintering temperatures, cheaper surface finishing methods (diamond free) and higher reliability of the components cut costs and therefore are crucial subjects of study. In addition to cost problems holding back progress are gaps in knowledge and experience and even unfounded caution on the part of application engineers. Sound internationally accepted standards would be helpful in overcoming these obstacles.

Si_3N_4 ceramics have developed into a family of very well established materials with many useful properties, of which resistance to thermal shock, high temperature stability, hardness and wear resistance are the most important. The high potential of this versatile class of ceramics is evident, but is far from being exhausted.

12
References

1. Nittler LR, Hoppe P, Alexander CMOD, Amari S, Eberhardt P, Gao X, Lewis RS, Strebel R, Walker RM, Zinner E (1995) The Astrophysical J 453: L25
2. Lee MR, Russel SS, Arden JW, Pillinger CT (1995) Meteoritics 30: 387
3. Sainte-Claire Deville H, Wöhler F (1859) Ann Chem Pharm 34: 248
4. Melner H (1896) German Patent No 88999
5. Weiss L, Engelhardt T (1910) Z Anorg Chem 65: 38
6. Wöhler L (1926) Z Electrochem 32: 420
7. Schröder F (ed) (1989) Gmelin Handbook of Inorganic and Organometallic Chemistry, 8th edn., Silicon, Suppl Vol B4, Springer, Berlin, Heidelberg, New York
8. DP 235 421 (1908), DP 238450 (1909)
9. Collins JF, Gerby RW (1955) J Met 7: 612
10. Reinhardt F (1958) Glas-Email-Keramo Technologie 9: 327
11. Parr NL, Martin GF, May ERW (1960) Preparation, Microstructure and Mechanical Properties of Silicon Nitride. In: Popper P (ed) Special Ceramics. Heywood, London, p 102
12. Popper P, Ruddlesden SN (1961) Trans Br Ceram Soc 60: 603
13. Deeley GG, Herbert JM, Moore NC (1961) Powder Metall 8: 145
14. Riley FL (2000) J Am Ceram Soc 83: 245
15. Herrmann M, Klemm H, Schubert C (2000) Silicon Nitride Based Hard Materials. In: Riedel R (ed) Handbook of Ceramic Hard Materials. Wiley-VCH, Weinheim, p 747

16. Chen IW, Becher PF, Mitomo M, Petzow G, Yen TS (1993) Symp Proc, MRS, Pittsburgh
17. Hampshire S (1994) Nitride Ceramics. In: Swain MV (ed) Structure and properties of ceramics. Cahn RW, Haasen P, Kramer EJ (eds) Materials Science and Technology, Comprehensive Treatment. VCH, Weinheim, p 119
18. Gmelin Handbook of Inorganic and Organometallic Chemistry, 8th edn. Springer, Berlin, Heidelberg, New York; (a) Schröder F (ed) (1996) Silicon Nitride: Mechanical and Thermal Properties: Diffusion. Silicon Suppl Vol B5b1; (b) Schröder F (ed) (1997) Silicon Nitride: Electronic Structure; Electrical, Magnetic and Optical Properties; Spectra; Analysis. Silicon Suppl Vol B5b2; (c) Schröder F (ed) (1991) Silicon Nitride in Microelectronics and Solar Cells. Silicon Suppl Vol B5c; (d) Schröder F (ed) (1995) Silicon Nitride: Electrochemical Behavior; Colloidal Chemistry and Chemical Reactions. Silicon Suppl Vol B5d1; (e) Schröder F (ed) (1995) Silicon Nitride: Chemical Reactions (continued). Silicon Suppl Vol B5d2; (f) Schröder F (ed) (1994) Non-Electronic Applications of Silicon Nitride. SiN_x. SiN_x:H. Silicon Suppl Vol B5e
19. Hoffmann MJ, Petzow G (eds) (1994) Tailoring of Mechanical Properties of Si_3N_4 Ceramics. NATO ASI Ser E 276. Kluwer Acad Publ, Dordrecht
20. Hoffmann MJ, Becher PF, Petzow G (eds) (1994) Silicon Nitride 93 – Proc Int Conf on Silicon Nitride-Based Ceramics, Stuttgart Trans Tech Publications Ltd, Aedermannsdorf, Switzerland
21. Lange H, Wötting G, Winter G (1991) Angew Chem Int Ed Engl 30: 1579
22. Wang CM, Pan X, Rühle M, Riley FL, Mitomo M (1996) J Mat Sci 31: 5281
23. Zerr A, Miehe G, Serghiou G, Schwarz M, Kroke E, Riedel R, Fueß H, Kroll P, Boehler R (1999) Nature 400: 340
24. Toraya H (2000) J Appl Cryst 33: 95
25. Grün R (1979) Acta Cryst B35: 800
26. Schwarz M, Miehe G, Zerr A, Kroke E, Poe BT, Fuess H, Riedel R (2000) Adv Mat 12: 883
27. Hiraga K, Tsuno K, Shindo D, Hirabayashi M, Hagashi S, Hirai T (1983) Phil Mag A 47: 483
28. Wendel JA, Goddard III WA (1992) J Chem Phys 97: 5048
29. He H, Sekine T, Kobayashi T, Hirosaki H, Suzuki J (2000) Phys Rev B62/17: 1
30. Kohatsu I, Mc Cauley JW (1974) Mat Res Bull 9: 917
31. Suematsu H, Petromic JJ, Mitchell TE (1996) Mat Sci Eng 209: 97
32. Chakraborty D, Mukerji J (1983) Mat Res Bull 17: 843
33. Sun EY, Becher PF, Plucknett KP, Hueh CH, Alexander KB, Waters SB (1998) J Am Ceram Soc 81: 2821 and 2831
34. Dusza J, Eschner T, Rundgren K (1997) J Mat Sci Lett 16: 1664
35. Ching W, Xu Y, Gale J, Rühle M (1998) J Am Ceram Soc 81: 3189
36. Hay JC, Sun EY, Pharr GM, Becher PF, Alexander KB (1998) J Am Ceram Soc 81: 2661
37. O'Hare PAG, Tomaszkiewicz I, Beck II CM, Seifert HJ (1999) J Chem Thermodynamics 31: 303
38. Liang JL, Topor L, Navrotsky A (1999) J Mat Res 14: 1959
39. Andrievskii RA, Spivak II (1984) Nitrid Kremnija i Materialy na ego ocnove, Metallurgija, Moskau, p 19
40. Kitayama M, Hirao K, Toriyama M, Kanzaki S (1999) J Am Ceram Soc 82: 3105
41. Watari K (2001) J Ceram Soc Jpn 109: S7
42. Kunz KP, Sarin VK, Davis RF, Bryan SR (1988) Mat Sci Eng A 105/106: 47
43. Kijima K, Shirasaki S (1976) J Chem Phys 65: 2668
44. Clancy WP (1974) Microscope 22: 279
45. Wild S, Grieveson P, Jack KH (1972) Special Ceramics 5: 385
46. Konstanovskii AV, Evseev AV (1994) High Temp (translated from Teplofizika Vysokikh Temperatur, Russia) 32: 25
47. Cerenius Y (1999) J Am Ceram Soc 82: 380
48. Schneider J, Frey F, Johnson N, Laschke K (1994) Z Kristallographie 209: 328
49. Kitayama M (1999) J Am Ceram Soc 82: 3263

50. Thompson DP (1993) New Grain Boundary Phases for Nitrogen Ceramics. In: Chen IW, Becher PF, Mitomo M, Petzow G, Yen TS (eds) Silicon Nitride Ceramics. Mat Res Soc Symp Proc 287: 79

51. Milhet X, Demenet JL, Rabier (1999) Phil Magaz Lett 79: 19

52. Wang CM, Riley FL (1996) J Eur Ceram Soc 16: 679.

53. Bowen LJ, Carruthers TG, Brook RJ (1978) J Am Ceram Soc 61: 335

53 a. Saito T, Iwamoto Y, Ukyo Y, Ikuhara Y (2000) Interface Characterisation of α-β Phase Transformation in Si_3N_4 by Transmission Electron Microscopy. In: Sakuma T, Sheppart LM, Ikuhara Y (eds) Grain Boundary Engineering in Ceramics, Am Ceram Soc, Westerville OH, p 173

54. Colquhoun J (1973) Proc Br Cer Soc 22: 207

55. Chase MW JR, Davies CA, Downey JR, Frurip DJ, McDonald RA, Syverud AN (1985) J Phys Chem Ref Data 14 Suppl. 1

56. The American Ceramic Society (ACerS) (ed) (1996) Cumulative Indexes of "Phase Equilibria Diagrams – Phase Diagrams for Ceramists". Vol I-XII and Annuals 91–93, Westerville, Ohio

57. Alper AM (ed) (1995) Phase Diagrams in Advanced Ceramics. Academic Press, London

58. Massalski TB (ed) (1990) Binary Alloy Phase Diagrams. American Society for Materials (ASM), Materials Park, Ohio

59. Petzow G, Effenberg G, Aldinger F (eds) (1988–2000) Ternary Alloys – A Comprehensive Compendium of Evaluated Constitutional Data and Phase Diagrams Materials Science International Series (MSI),Vol. 1–17, Stuttgart

60. Blegen K (1976) PhD Thesis, Techn University of Trondheim, Norway

61. Hillert M, Jonsson S, Sundman B (1992) Z Metallkde. 83: 648

62. Heuer AH, Lou VL (1990) J Am Ceram Soc 73: 2785

63. Hallstedt B (1992) CALPHAD 16: 53

64. Swamy V, Saxana SK, Sundman B (1994) J Geoph Res 99: 11 and 787

65. Wriedt HA (1990) Bull Alloy Phase Diagr 11: 43

66. Rocabois P, Chatillon C, Bernard C (1996) J Am Ceram Soc 79: 1351 and 1361

67. Weiss J, Lukas HL, Petzow G (1983) Calculation of Phase Equilibria in Systems based on Si_3N_4. In: Riley FL (ed) Progress in Nitrogen Ceramics. NATO ASI Series E. 65: 77

68. Richter HJ, Herrmann M, Hermel W (1991) J Eur Ceram Soc 7: 3

69. Lukas HL, Weiss J, Krieg H, Henig ET, Petzow G (1982) High Temp High Press 14: 607

70. Gauckler LJ (1976) PhD Thesis, University of Stuttgart

71. Jack KH (1976) J Mat Sci 11: 1135

72. Gauckler LJ, Lukas HL, Petzow G (1975) J Am Ceram Soc 58: 346

73. Lumby RJ, Noith B, Taylor AJ (1975) Spec Ceram Br Ceram Res Association 6: 283

74. Naik JK, Gauckler LJ, Tien TY (1978) J Am Ceram Soc 61: 332

75. Bergmann B, Ekström T, Micski A (1991) J Eur Ceram Soc 8: 141

76. Boskovic S, Gauckler LJ, Petzow G, Tien TY (1977) Powder Metal Intern 9: 185; (1978) 10: 184; (1979) 11: 169

77. Hillert M, Jonsson S (1992) Z Metallkde 83: 720

78. Dumitrescu LFS, Sundman B (1995) J Eur Ceram Soc 15: 89 and 239

79. Anya CC, Hendrix A (1992) J Eur Ceram Soc 10: 65

80. Sekercioglu J, Wills RR (1979) J Am Ceram Soc 62: 590

81. Hillert M, Jonsson S (1992) CALPHAD 16: 199

82. Tien TY, Petzow G, Gauckler LJ, Weiss J (1983) Phase Equilibrium Studies in Si_3N_4-Metal Oxides Systems. In: Riley FL (ed) Progress in Nitrogen Ceramics. NATO ASI Ser E 65. Kluwer Acad Publ, Dordrecht, p 89

83. Thompson DP, Korgul P, Hendry A (1983) The structural characterisation of SiAlON Polytypoids. In: Riley FL (ed) Progress in Nitrogen Ceramics. NATO ASI Ser E 65. Kluwer Acad Publ, Dordrecht, p 61

84. Gauckler LJ, Weiss J, Tien TY, Petzow G (1978) J Am Ceram Soc 61: 397

85. Zhou Y, Vleugels J, Laoui T, Ratchev P, Van Der Biest O (1995) J Mater Sci 30: 4584

86. Liang JJ, Navrotzky A, Leppert VJ, Paskowitz MJ, Risbud SH, Ludwig T, Seifert HJ, Aldinger F (1999) J Mater Res 14: 4630
87. Mitomo M, Kuramato N, Tsutsumi M, Suzuki H (1987) Yogyo Kyokaishi 86: 880
88. Ekström T, Käll PO, Nygren M, Olsson PO (1989) J Mater Sci 24: 1853
89. Huseby C, Lukas HL, Petzow G (1975) J Am Ceram Soc 58: 337
90. Thompson DP, Gauckler LJ (1977) J Am Ceram Soc 60: 470
91. Kim DJ, Greil P, Petzow G (1987) Adv Ceram Mat 2: 817; 2: 822
92. Ekelund M, Forslund M, Erikson G, Johansson T (1988) J Am Ceram Soc 71: 956
93. Neidhardt V, Schubert H, Bischoff E, Petzow G (1994) Key Eng Mat 89: 187
94. Wada H, Wang HJ, Tien TY (1988) J Am Ceram Soc 71: 837
95. Jha A (1993) J Mat Sci 28: 3069
96. Wang MJ, Wada H (1990) J Mat Sci 25: 1690
97. Lange FF (1980) J Am Ceram Soc Bull 59: 249
98. Hampshire S, Park HK, Thompson DP, Jack KH (1978) Nature 274: 880
99. Huang ZK, Chen IW (1996) J Am Ceram Soc 79: 2091
100. Huang ZK, Tien TY, Yen TS (1986) J Am Ceram Soc 69: C241
101. Kaiser A (1999) PhD Thesis, RWTH Aachen
102. Mitomo M, Izumi F, Horiuchi S, Matzui J (1982) J Mater Sci 17: 2359
103. Mueller R (1981) PhD Thesis, University of Stuttgart
104. Lange FF (1978) J Am Ceram Soc 61: 53
105. Clarke DR, Lange FF (1980) J Am Ceram Soc 63: 585
106. Slasor S (1986) PhD Thesis, University of Newcastle Upon Tyne
107. Slasor S, Liddell K, Thompson DP (1986) The Role of Nd_2O_3 as an Additive in the Formation of α'- and β'-Sialons. In: Howlett SP, Taylor D (eds) Proc Special Ceramics, p 35; (1986) Br Ceram Proc 37: 61
108. Hohnke H, Gauckler LJ, Schneider G, Tien TY (1979) Am Ceram Soc Bull 58: 885
109. Jack KH (1978) Mat Sci Res 11: 561
110. Gauckler LJ, Hohnke H, Tien TY (1980) J Am Ceram Soc 63: 35
111. Schubert H, Gehrke E (1994) Processing, Phase Formation and Creep Behavior of Si_3N_4 with Y_2O_3 and Al_2O_3. In: Hoffmann MJ, Petzow G (eds) Tailoring of Mechanical Properties of Si_3N_4 Ceramics. NATO ASI Ser E 276, Kluwer Acad, Dordrecht, p 245
112. Sun WY, Huang ZK, Tien TY, Yan DS (1991) Mat Lett 11: 67
113. Hoffmann MJ (1994) High Temperature Properties of Yb-Containing Si_3N_4. In: Hoffmann MJ, Petzow G (eds) Tailoring of Mechanical Properties of Si_3N_4 Ceramics, Kluwer Acad Publ, Dordrecht, p 233
114. Weiss J, Gauckler LJ, Tien TY (1979) J Am Ceram Soc 62: 632
115. Hampp E (1993) PhD Thesis, University of Stuttgart
116. Nishimura T, Mitomo M (1995) J Mater Res 10: 240
117. Huang ZK, Sun WY, Yen TS (1985) J Mat Sci Lett. 4: 255
118. Hewett CL, Cheng YB, Muddle BC, Trigg MB (1998) J Eur Ceram Soc 18: 417
119. Hewett CL, Cheng YB, Muddle BC, Trigg MB (1998) J Am Ceram Soc 81: 1781
120. Wood CA, Cheng YB (2000) J Eur Ceram Soc 20: 357
121. Wang PL, Zhang C, Sun WY, Yan DS (1999) J Eur Ceram Soc 19: 553
122. Wood CA, Zhao H, Cheng YB (1999) J Am Ceram Soc 82: 421
123. AS3. Ekström T (1997) α-SiAlON and α/β-SiAlON composites; Recent research. In: Babini GN, Haviar M, Sajgalik P (eds) Engineering Ceramics'96; Higher Reliability through Processing, Kluwer Academic Publications, The Netherlands, p 147
124. Ekström T, Nygren M (1992) J Am Ceram Soc 75: 259
125. Nordberg LO, Shen Z, Nygren M, Ekström T (1997) J Eur Ceram Soc 17: 575
126. Sun WY, Yan DS, Gao L, Mandal H, Liddel K, Thompson DP (1995) J Eur Ceram Soc 15: 349
127. Nordberg LO (1997) PhD Thesis, University of Stockholm
128. Mandal H, Thompson DP (1999) J Eur Ceram Soc 19: 543
129. Mukerji J, Das PK, Greil P, Petzow G (1987) Ceram Intern 13: 215
130. Mandal H, Thompson DP (1996) J Mater Sci Letters 15: 1435

131. Mandal H, Hoffmann MJ (2000) Novel Developments in α-SiAlON Ceramics. In: Sajgalik P, Lences Z (eds) Engineering Ceramics: Multifunctional Properties–New Perspectives. Trans Tech Publications Ltd, Switzerland, p 131
132. Izhevsky VA, Genova LA, Bressiani JC, Aldinger F (2000) J Eur Ceram Soc 20: 2275
133. Wang PL, Tu HY, Sun WY, Yan DS, Nygren M, Ekström T (1995) J Eur Ceram Soc 15: 689
134. Zhang C, Sun WY, Yan DS (1998) J Mater Sci Letters 17: 583; (1999) J Eur Ceram Soc 19: 33
135. Mitomo M, Ishida A (1999) J Eur Ceram Soc 19: 7
136. Sun WY, Tu HY, Wang PL, Yan DS (1997) J Eur Ceram Soc 17: 789
137. Liddell K, Thompson DP, Wang PL, Sun WY, Gao L, Yan DS (1998) J Eur Ceram Soc 18: 1479
138. Mandal H, Thompson DP, Ekström T (1993) J Eur Ceram Soc 12: 421
139. Lewis MH, Jumali MHH, Lumby RJ (1997) Nd- and Gd-doped α'/β' SiAlON Ceramics. In: Niihara K, Hirano S, Kanzaki S, Komeya K, Moriuaga K (eds) Ceramic Materials and Components for Engines. Japan Fine Ceramics Association, Tokyo, p 643
140. Zhang E, Liddell K, Thompson DP (1996) Brit Ceram Trans 95: 169
141. Yu ZB, Thompson DP, Bhattit R (2000) J Eur Ceram Soc 20: 1815
142. Nunn SD, Hohnke H, Gauckler LJ, Tien TY (1978) J Am Ceram Soc Bull 57: 321
143. Weiss J (1977) MS thesis, University of Stuttgart
144. Chee KS, Cheng YB, Smith ME (1995) J Eur Ceram Soc 15: 1213
145. Sun WY, Yan DS, Gao L, Mandal H, Lidell K, Thompson DP (1996) J Eur Ceram Soc 16: 1277
146. Kaiser A, Telle R, Richter HJ, Herrmann M, Hermel W (1996) Konstitutionsuntersuchungen im System Nd-Si-Al-ON. In: Petzow G, Tobolski J, Telle R (eds) Hochleistungs-Keramiken. VCH, Weinheim, p 627; (2001) Z. Metallkd. 92: 1163
147. Rosenflanz A, Chen IW (1999) J Am Ceram Soc 82: 1025
148. Shen Z, Nygren M (1997) J Eur Ceram Soc 17: 1639
149. Shen Z, Nygren M (2000) Nd-Doped α-Sialon and Related Phases: Stability and Compatibility. In: Hampshire S, Pomeroy MJ (eds) Nitrides and Oxynitrides. Trans Tech Publications Ltd, Switzerland, p 191
150. Nordberg LO, Nygren M, Käll PO, Shen Z (1998) J Am Ceram Soc 81: 1461
151. Shen Z, Ekström T, Nygren M (1996) J Am Ceram Soc 79: 721
152. Herrmann M, Kurama S, Mandal H (2002) J Eur Ceram Soc 22: 109
153. Zhao R, Cheng YB (1995) J Eur Ceram Soc 15: 1221
154. Chen WW, Li YW, Sun WY, Yan DS (2000) J Eur Ceram Soc 20: 1327
155. Sun WY, Tien TY, Yen TS (1991) J Am Ceram Soc 74: 2547 and 2753
156. Wisnudel M (1991) MS Thesis, University of Michigan, Ann Arbor
157. Hwang CM (1988) PhD Thesis, University of Michigan, Ann Arbor
158. Naik JK, Tien TY (1979) J Am Ceram Soc 62: 642
159. Thompson DP (1986) Phase relationships in Y-Si-Al-O-N ceramics. In: Tressler RE, Messing GL, Pantano CG, Newnham RE (eds) Mat Sci Res Proc 21st Univ Conf in Ceramic Science; Tailoring Multiphase and Composite Ceramics 20. Plenum Press, New York, p 79
160. Sun WY, Huang ZK, Chen JX (1983) Trans J Brit Ceram Soc 82: 173
161. Huang ZK, Greil P, Petzow G (1983) J Am Ceram Soc 66: C96
162. Huang ZK, Tien TY (1996) J Am Ceram Soc 79: 1717
163. Wisnudel M, Tien TY (1994) J Am Ceram Soc 77: 2653
164. Klemm H, Herrmann M, Schubert C, Hermel W (1995/1996) High Temp–High Press 27/28: 449
165. Redington Keely W, Redington M, Hampshire S (2000) Liquid Formation in the Y-Si-Al-O-N System. In: Hampshire S, Pomeroy MJ (eds) Nitrides and Oxinitrides. Mater Sci Forum 325–326. Trans Tech Publications Ltd, Switzerland, p 237
166. Bandyopadhyay S, Hoffmann MJ, Petzow G (1996) J Am Ceram Soc 79: 1537
167. Ekström T, McKenzie KJD, Ryan MJ, Brown WM, White GV (1997) J Mater Chem 7: 505

168. Kolitsch V, Seifert HJ, Ludwig T, Aldinger F (1999) J Mater Res 14: 447
169. Stutz D, Greil P, Petzow G (1986) J Mat Sci Lett 5: 335
170. Izumi F, Mitomo M, Suzuki J (1982) J Mater Sci Lett 1: 533
171. Shen Z, Ekström T, Nygren M (1996) J Phys D Appl Phys 29: 893
172. Weiss J (1980) PhD Thesis, University of Stuttgart
173. Seifert HJ (1999) Z Metallkd 90: 1016
174. Lange FF (1979) J Am Ceram Soc 62: 617
175. Cao GZ, Huang ZK, Yan DS (1989) Sci China Ser A 32: 429
176. Mahoney FM (1992) PhD thesis, University of Stuttgart
177. Dörner P (1982) PhD Thesis, University of Stuttgart
178. Wang PL, Li YW, Yan DS (2000) J Eur Ceram Soc 20: 1333
179. Gauckler LJ, Hucke E, Lukas HL, Petzow G (1978) CALPHAD VII: 49; (1979) J Mat Sci 14: 1513
180. Ran Q (1987) PhD Thesis, University of Stuttgart
181. Gauckler LJ, Petzow G (1977) Representation of Multicomponent Silicon Nitride Based Systems. In: Riley FL (ed) Nitrogen Ceramics, Noordhoff-Leyden, p 41
182. Metselaar R, Yan DS (1999) Pure Appl Chem 71: 1765
183. Oyama Y, Kamigaito O (1971) Jap J Appl Phys 10/11: 1637
184. Jack KH, Wilson WJ (1972) Nature (London), Phys Sci 283: 28
185. Willems HX, Hendrix MMRM, Metselaar R, de With G (1992) J Eur Ceram Soc 10: 327; 10: 339
186. Hallstedt B, Hillert M, Selby M, Sundman B (1994) CALPHAD 18: 31
186a. Sekine T, He H, Kobayashi T, Tansho M, Kimoto K (2001) Chem Phys Let 344: 395
187. Cao GZ, Metselaar R (1991) Chem Mater 3: 242
188. Mandal H, Thompson DP, Jack KH (1999) Key Eng Mat 154–160: 1
189. Mandal H, Hoffmann MJ (1999) J Am Ceram Soc 82: 229
190. Lathrop D (2000) Ceram Bull 79: 54
191. Schwier G, Nietfeld G, Franz G (1989) Mat Sci Forum 47: 1
192. Herrmann M, Schulz I, Hintermayer J (1995) Materials From Low Cost Silicon Nitride Powders In: Galassi C (ed) Proc 4th EcerS Conf, Riccione, Gruppo Editoriale Faenza Editrice, p 211
193. Chang FW, Liou TH, Tsai FM (2000) Thermochimica Acta 354: 71
194. Liu YD, Kimura S (1999) Powder technology 106: 160
195. Hirotsuru H, Isozaki K, Yoshida A (1993) Trans Mat Res Soc Jap 14A: 815
196. Nakamura M, Kuranari Y, Imamura Y (1987) Characterisation and synthetic process of Si_3N_4 material powders. In: Somiya S, Mitomo M, Yoshimura M (eds) Silicon Nitride I. Elsevier, London, p 40
197. Kohtoku Y (1987) Developments in Si_3N_4 powder prepared by the imide decomposition method. In: Somiya S, Mitomo M, Yoshimura M (eds) Silicon Nitride I. Elsevier, London, p 71
198. Cochran GA, Conner CL, Eismann GA, Weimer AW, Carroll DF, Dunmead SD, Hwang CJ (1994) The Synthesis of a High Quality, Low Cost Silicon Nitride Powder by the Carbothermal Reduction of Silica. In: Hoffmann MJ, Becher PF, Petzow G (eds) Silicon Nitride 93. Key Eng Mater 89–91. Trans Tech Publications Ltd, Switzerland, p 3
199. Bandyopadhyay S, Mukerji J (1991) Ceram Inter 17: 171
200. Ishii T, Sano A, Imai I (1987) α-Si_3N_4 powder produced by nitriding silica using carbothermal reduction. In: Somiya S, Mitomo M, Yoshimura M (eds) Silicon Nitride I. Elsevier, London, p 59
201. Bandyopadhyay S, Mukerji J (1992) Ceram Inter 18: 308
202. Arik H, Saritas S, Gunduz M (1999) J Mat Sci 34: 835
203. Rahman A, Riley FL (1989) J Eur Ceram Soc 5: 11
204. Mukerji J, Bandyopadhyay S (1988) Adv Ceram Mat 3: 369
205. Brink R, Woditsch P (1998) cfi/Ber DKG, 75/7: 22
206. Dorn FW, Krause W, Schröder F (1992) Adv Mat 4: 221

207. Jennett TA, Harmsworth PD, Jones AG (1994) Ultra fine crystalline Silicon Nitride from a continuous Gas Phase Plasma Route. In: Hoffmann MJ, Becher PF, Petzow G (eds) Silicon Nitride 93. Key Eng Mat 89–91. Trans Tech Publications Ltd, Switzerland, p 47
208. Grabis J, Zalite I, Reichel U (2000) cfi/Ber DKG 77/7: 9
209. Kubo N, Futaki S, Shiraishi K (1987) Synthesis of ultrafine Si_3N_4 powder using the plasma process and powder characterisation. In: Somiya S, Mitomo M, Yoshimura M (eds) Silicon Nitride I. Elsevier, London, p 93
210. Friedrich M, Mohr R, Drost H, Mach R, Gey E (1997) Silicates Industriels 11: 1–2
211. Drost H, Friedrich M, Mohr R, Gey E (1997) Nucl Instr and Methods in Physics Res 122: 598
212. Borsella E, Caneve L, Fantoni R, Piccirillo S, Basili N, Enzo S (1989) Appl Surface Sci 36: 213
213. Kroke E, Li YL, Konetschny C, Lecomte E, Fasel C, Riedel A (2000) Mat Sci Eng R26: 97
214. Merzhanov AG (1995) Ceram Int 21: 371
215. Temer MR, Swenser SP, Cheng YB (2001) Characterization of Multi-Cation stabilized Alpha-SiAlON Materials. In: Heinrich JG, Aldinger F (eds) Ceramic Materials and Components for Engines, 7th International Symposium. Wiley VCH, Weinheim, p 447
216. Hirata T, Akiyama K, Morimoto T (2000). J Eur Ceram Soc 20: 1191
217. Bermudo VJ Osendi MI (1999) Ceram Intern 25: 607
218. Wu Y, Zhuang H, Wu F, Dollimore D, Zhang B, Chen-Li W (1998) J Mat Res 13: 166
219. Herrmann M, Schulz I, Hermel W, Schubert C, Wendt A (2001) Z Metallkde 92: 788
220. Arakawa T (1987) State of the art of silicon nitride powders obtained by thermal decomposition of $Si(NH)_2$ and the injection molding thereof. In: Somiya S, Mitomo M, Yoshimura M (eds) Silicon Nitride I. Elsevier, New York, p 81
221. Brink R, Lange H (1994) Investigations on the Synthesis of fine-grained, high-purity β-Si_3N_4 Powder by Crystallization of amorphous Precursors. In: Hoffmann MJ, Becher PF, Petzow G (eds) Silicon Nitride 93. Key Eng Mat 89–91. Trans Tech Publications Ltd, Switzerland, p 73
222. Borsella E, Botti S, Martelli S, Alexandrescu R, Cesile MC, Nesterenko A, Giorgi R, Turtu S, Zappa G (1997) Silicates Industriels 3: 1–2
223. Niihara K (1991) J Ceram Soc Jpn Int Edition 99: 945
224. Petzow G, Sersale R (1987) Pure & Appl Chem 59: 1673
225. Samsonov GV, Kulik OP, Poluschschick VS (1978) Poluschenie i Methodij analisa Nitridov. Naukowa Dumka, Kiev, p 273
226. Rabe T, Sontag E, Kranz G, Röhl K, Linke D (1990) Analytische Bestimmung der in Siliciumnitrid vorliegenden Phasen. In: Kriegesmann J (ed) Keramische Werkstoffe, Deutscher Wirtschaftsdienst, Köln, p 6.1.6.1
227. Haßler I, Förster O, Schwetz KA (2000) cfi/Ber DKG 77/7: D11
228. Peplinski B, Schultze D, Wenzel J (2001) Interlaboratory comparison (round robin) on the application of the Rietveld method to quantitative phase analysis by X-ray and neutron diffraction. In: Delhez R, Mittemeijer EJ (eds) Proc 7th European powder diffraction conference (EPDIC-7), Barcelona, Spain, 20th 23rd May 2000, Trans Tech Publications Ltd, Switzerland, p 124
229. Matern N, Riedel A, Wassermann A (1993) Mat Sci Forum 133: 39
230. Gazzara CP, Messier DR (1977) Ceram. Bull 56: 777
231. Käll PO, Ekström T (1990) J Eur Ceram Soc 6: 1191
232. Goeb O (2001) PhD Thesis, Technical University of Dresden
233. Bermudo J, Osendi MI, Fierro JLG (2000) Ceram Intern 26: 141
234. Hirao K, Tsuge A, Toriyama M, Kanzaki S (1999) J Am Ceram Soc 82: 3263
235. Kaiser G, Schubert H (1993) J Eur Ceram Soc 11: 253
236. Sunderkötter JD, Grallath E, Jenett M (1993) Fresenius J Anal Chem 346: 237
237. Peuckert M, Greil P (1987) J Mater Sci 22: 3717
238. Richter HJ, Herrmann M (1991) J Mat Sci Lett 10: 783

239. Berger LM (1993) Application of Adsorption for the Characterization of Solids in Ceramic and Related Technologies. In: Aldinger F (ed) PTM '93, DGM Informationsgesellschaft, Oberursel, p 677
240. Scarlett B (1996) Characterisation of Particles and powders. In: Brook RI (ed) Processing of Ceramics, Part I.VCH, Weinheim, p 99
241. Boden G, Nitsche R (1993) Sprechsaal 126: 220
242. Sonnefeld J (1996) Physicochem Eng Aspects 108: 27
243. Petzow G (1988) Pract Met 25: 53
244. Reed JS (1995) Principles of Ceramic Processing, John Wiley & Sons Inc, New York, p 137
245. Riedel G, Krüner H (1993) Si_3N_4 material with high strength (1400 MPa). In: Duran P, Fernandez JF (eds) Third Euro-Ceramics 3. Faenza Editrice Iberica, San Vicente, p 453
246. Yoshimura M, Nishioka T, Yamakawa A, Miyake M (1995) J Ceram Soc Jpn 103: 407
247. Brook RJ (ed) (1990) Processing of Ceramics, Part I and II, Wiley-VCH, Weinheim
248. Riedel G, Schubert J (1996) Keram Zeit 48: 294 and 396
249. Riedel G, Krieger S (1996) Keram Zeit 48: 192
250. Kosmac T, Novak S, Sajko M (1997) J Eur Ceram Soc 17: 427
251. Hackley VA (1997) J Am Ceram Soc 80: 2315
252. Schwelm M (1992) PhD Thesis, University of Stuttgart
253. Bergström L (1996) J Am Ceram Soc 79: 3033
254. Lange FF, Luther EP (1994) Colloidal Processing of Structurally Reliable Si_3N_4. In: Hoffmann MJ, Petzow G (eds) Tailoring of Mechanical Properties of Si_3N_4 Ceramics. NATO ASI Series E. Vol 276, Kluwer Acad Publ Dordrecht, p 3
255. Vieth S, Mitzner E, Finke B, Linke D (2000) cfi Ber DKG 77(10): E63
256. Kuzjukevies A Ishizaki K (1993) J Am Ceram Soc 76: 2373
257. Pugh RI, Bergström L (1994) Surface and Colloid Chemistry in Advanced Ceramics Processing. Marcel Decker Inc, New York
258. Wang L, Sigmund W, Aldinger F (2000) J Am Ceram Soc 83: 691 and 697
259. Negita K (1985) J Mater Sci Letters 4: 755
260. Kim J, Iseki T (1996) J Am Ceram Soc 79: 2744
261. Wang L, Roy S, Sigmund W, Aldinger F (1999) J Eur Ceram Soc 19: 61
262. Wang C, Riley FL (1992) J Eur Ceram Soc 10: 83
263. Luther EP, Lange FF, Pearson DS (1995) J Am Ceram Soc 78: 2009
264. Wang L, Sigmund W, Roy S, Aldinger F (1999) J Mater Res 14: 4562
265. Greil P (1982) PhD thesis, University of Stuttgart
266. Herrmann M, Putzky G, Siegel S, Hermel W (1992) cfi Ber DKG 69: 375
267. Yokoyama K, Wada S (2000) J Ceram Soc Jpn 108: 6
268. Neidhardt U (1993) PhD Thesis, University of Stuttgart
269. Riedel G, Bestgen H, Herrmann M (1998) cfi, Ber DKG, 75/12: 30
270. Grechkovich C, Prochazka S (1981) J Am Ceram Soc 64: C96
271. Messier DR, Patel PJ (1993) Chemically induced defects in oxinitride glasses. In: Chen IW, Becher PF, Mitomo M, Petzow G, Yen TS (eds) Silicon Nitride Ceramics, Mat Res Soc Symp 287; Mat Res Soc, Pittsburgh, p 365
272. Herrmann M, Goeb O (2001) J Eur Ceram Soc 21: 304 and 461
273. Yokoyama K, Wada S (1999) Effect of Sintering Atmosphere on the α/β-Transformation of Si_3N_4. In: Suzuki H, Komeya K, Uematsu K (eds) Novel Synthesis and Processing of Ceramics, Key Eng Mat 159–160. Trans Tech Publications, Switzerland, p 209
274. Yokoyama K, Wada S (2000) J Ceram Soc Jpn 108: 230 and 357
275. Herrmann M, Schubert C, Rendtel A, Hübner H (1998) J Am Ceram Soc 81: 1095
276. Pompe R, Carlson R (1983) Sintering of Si_3N_4 based Materials using the Powder Bed Technique. In: Riley FL (ed) Progress in Nitrogen Ceramics, NATO ASI Ser E. 65. Kluwer Academic Publishers, Dordrecht, p 219
277. Giachello A, Martinengo PC, Tommasini G, Popper P (1979) J Mat Sci 14: 2855

278. Warnecke G (ed) (2000) Zuverlässige Hochleistungskeramik. Universität Kaiserslautern, Kaiserslautern
279. Richerson DW (1992) Modern Ceramic Engineering. Marcel Decker Inc, New York
280. Pattimore J, Nishio A, Dewitte C, Unno Y, Masuda M (1997) Optimisation of Grinding for Cylindrical Silicon Nitride Components for Mass Production. In: Niihara K, Hirano S, Kanzaki S, Komeya K, Moriuaga K (eds) Ceramic Materials and Component for Engines. Japan Fine Ceramics Association, Tokyo, p 406
281. Greil P (1997) Near Net Shape Manufacturing of Polymer Derived Ceramics In: Baxter J, Cot L, Fordham R Gabis V, Hellot Y (eds) Euro Ceramics 5 Part 3- Key Eng Mat 132-136 ; Trans Tech Publications, Switzerland, p 1981
282. Lange FF (1979) J Am Ceram Soc 62: 428
283. Krämer M, Hoffmann MJ, Petzow G (1993) Acta metall mater 41: 2939
284. Krämer M, Hoffmann MJ, Petzow G (1993) J Am Ceram Soc 76: 2778
285. Dressler W (1993) PhD thesis, University of Stuttgart
286. Chen IW, Hwang SL (1993) Superplastic SiAlON–A Birds Eye View of Silicon Nitride Ceramics. In: Chen IW, Becher PF, Mitomo M, Petzow G, Yen TS (eds) Silicon Nitride Ceramics, Mat Res Soc Symp 287; Mat Res Soc, Pittsburgh, p 209
287. Bestgen H, Boberski C (1998) HTEM Analyse der Mikrostrukturentwicklung während des Flüssigphasensinterns eines Si_3N_4/HfO_2 Composites In: Heinrich J, Ziegler G, Hermel W, Riedel H (eds) Werkstoffwoche'98 Band VII, Symposium 9, Keramik, Wiley-VCH, Weinheim, p 401
288. Hwang SL, Chen IW (1994) J Am Ceram Soc 77: 1711
289. Mitomo M, Hirotsuru H, Suematsu H, Nishimura T (1995) J Am Ceram Soc 78: 211
290. Herrmann M, Schulz I, Schubert C, Zalite I, Ziegler G (1998) cfi/Ber DKG 75: 38
291. Okamoto Y, Hirosaki N, Akimüne Y, Mitomo M (1997) J Ceram Soc Jpn 105: 476
292. Kitayama M, Hirao K, Toriyama M, Kanzaki S (1998) Anisotropie Ostwald Ripening in β-Si_3N_4 with different Lanthanide Additives. In: Messing GL, Hirano S, Lange FF (eds) Ceramic Processing Science. Ceram Transactions Ser 83: 517; (1999) J Ceram Soc Jpn 107: 930 and 107: 995
293. Becher PF, Sun EY, Plucknett KP, Alexander KB, Hsuch CH, Lin HAT, Waters SB, Westmoreland CG (1998) J Am Ceram Soc 81: 2821
294. Woetting G, Feuer H, Gugel E (1993) The Influence of Powders and Processing Methods on Microstructure and Properties of Dense Silicon Nitride. In: Chen IW, Becher PF, Mitomo M, Petzow G, Yen TS (eds) Silicon Nitride Ceramics, Mat Res Soc Symp Proc 287. Mat Res Soc, Pittsburgh, p 133
295. Dressler W, Kleebe HJ, Hoffmann MJ, Rüühle M, Petzow G (1996) J Eur Ceram Soc 16: 3
296. Kessler S, Herrmann M, Schubert C (1991) Mat Sci Forum 94-96: 821
297. Tien TY (1993) Silicon Nitride Ceramics–Alloy Design. In: Chen IW, Becher PF, Mitomo M, Petzow G, Yen TS (eds) Silicon Nitride Ceramics. Mat Res Soc Symp 287. Mat Res Soc, Pittsburgh, p 51
298. Wang LL, Tien TY, Chen IW (1998) J Am Ceram Soc 81: 2677
299. Emoto H, Mitomo M (1997) J Eur Ceram Soc 17: 797
300. Kanamaru M (1994) PhD Thesis, University of Stuttgart
301. Riedel G, Bestgen H, Herrmann M (1999) cfi/Ber DKG 71 No 1: 24
302. Björklund H, Falk LKL, Rundgren K, Wasen J (1997) J Eur Ceram Soc 17: 1285
303. Pyzik AJ, Beaman DR (1993) J Am Ceram Soc 76: 2737
304. Pyzik AJ, Carrol DF (1994) Annu Rev Mater Sci 24: 168
305. Ziegler G, Lehner W, Kleebe HJ (1999) Br Ceram Proc 60: 5
306. Bränvall P, Rundgren K (1999) Br Ceram Proc 60: 7
307. Kim HD, Han BD, Park DS (2001) Process to obtain Bimodal Microstructure in Silicon Nitride. Paper presented at 25th Annual Conference on Composites, Advanced Ceramics, Materials, and Structures. Am Ceram Soc, Cocoa Beach, Fl
308. Lai KR, Tien TY (1993) J Am Ceram Soc 76: 91
309. Mitomo M, Uenosono J, (1992) J Am Ceram Soc 75: 103

310. Mitomo M (1999) In-Situ Microstructural Control in Engineering Ceramics. In: Niihara K, Sekino T, Yasuda E, Sasa T (eds) The Science of Engineering Ceramics II. Key Eng Mat 161–163. Trans Tech Publications, Switzerland, p 53
311. Park DS, Lee SY, Kim HD, Yoo BJ, Kim BA (1998) J Am Ceram Soc 81: 1876
312. Stemmer S, Roebben G, Van der Biest O (1998) Acta mater 46: 5599
313. Cinibulk MK, Kleebe HJ Schneider GA, Rühle M (1993) J Am Ceram Soc 76: 2081
314. Wang CM, Pan X, Hoffmann MJ, Cannon RM, Rühle M (1996) J Am Ceram Soc 79: 788
315. Lee CJ, Kim DJ, Kang ES (1999) J Am Ceram Soc 82: 753
316. Bando Y, Mitomo M, Kurashima K (1998) J Mater Synthesis and Processing 6: 359
317. Kleebe HJ (1997) J Ceram Soc Jpn 105: 453
318. Clarke DR (1987) J Am Ceram Soc 70: 15
319. Raj R, Lange FF (1981) Acta Met 29: 1993
320. Yoshiya M, Tanaka I, Adachi H (2000) Atomic Structure and Chemical Bonding of Intergranular film in Si_3N_4-SiO_2 Ceramics. In: Sajgalik P, Lences Z (eds) Engineering Ceramics: Multifunctional Properties-New Perspectives. Trans Tech Publications Ltd, Switzerland, p 107
321. Lewis MH (1994) Crystallisation of Grain Boundary Phases in Silicon Nitride and SiAlON Ceramics. In: Hoffmann MJ, Petzow G (eds) Tailoring of Mechanical Properties of Si_3N_4 Ceramics. NATO ASi Ser E 276, Klüwer Academic Publishers, Dordrecht, p 217
322. Kleebe HJ, Hoffmann MJ, Rühle M (1992) Z Metallkde 83: 610
323. Okamoto K, Kleebe HJ, Ota K, Pezzotti G (1999) J Jap Inst Mat 63: 1479
324. Tanaka J, Kleebe HJ, Cinibulk MK, Bruley J, Clarke DR, Rühle M (1994) J Am Ceram Soc 77: 911
325. Kleebe HJ, Cinibulk MK, Cannon RM, Rühle M (1993) J Am Ceram Soc 76: 1969
326. Björklund H, Falk LKL (1997) J Eur Ceram Soc 17: 1301
327. Bobeth M, Clarke DR, Pompe W (1999) J Am Ceram Soc 82: 1537
328. Ackler HD, Chaing YM (1997) J Am Ceram Soc 80: 189
329. Keblinski P, Philpert SR, Wolf D, Gleiter H (1996) Phys Rev Lett 77: 2965
330. Golczewski JA, Seifert HJ, Aldinger F (2001) Z Metallkde 92: 695
331. Menon M, Chen IW (1995) J Am Ceram Soc 78: 545 and 553
332. Mandel H, Hoffmann MJ (1999) Br Ceram Proc 60: 11
333. Chen IW, Rosenflanz A (1997) Nature 389: 701
334. Klemm H, Herrmann M, Reich T, Schubert C, Frassek L, Wötting G, Gugel E, Nietfeld G (1998) J Am Ceram Soc 81: 1141
335. Mandal H (1999) J Eur Ceram Soc 19: 2349
336. Sheu TS (1994) J Am Ceram Soc 77: 2345
337. Rosenflanz A (1999) Current Opinion in Sol State & Mat Sci 4: 453
338. Kim J, Rosenflanz A, Chen IW (2000) J Am Ceram Soc 83: 1819
339. Wang CM, Hirosaki N, Mitomo M (2000) J Ceram Soc Jpn 108: 298
340. Herrmann M (1998) annual report 1998 Fraunhofer IKTS, Dresden
341. Kurama S, Herrmann M, Mandal H (2001) J Eur Ceram Soc accepted for publication
342. Rosenflanz A, Chen IW (1999) J Eur Ceram Soc 19: 2337
343. Mandal H. Thompsson DP, Ekström T (1993) J Eur Ceram Soc 12: 421
344. Cao GZ, Melselaar R, Ziegler G (1992) Microstructure and properties of mixed $\alpha'+\beta'$-Sialons. In: Carlsson R, Johansson T, Kahlman L (eds) Ceramics Materials and Components for Engines. Elsevier, London, p 188
345. Täffner U, Carle V, Schäfer U, Predel F, Petzow G (1991) Pract Met 28: 592
346. Mücklich F, Ohser J, Hartmann S, Hoffmann MJ, Petzow G (1994) 3D-Characterisation of Sintered Microstructures with Prismatic Grains–A Precondition for Microstructural Modelling of Si_3N_4 Ceramics. In: Hoffmann MJ, Petzow G (eds) Tailoring of Mechanical Properties of Si_3N_4 Ceramics. NATO ASI Ser E 276. Kluwer Academic Publishers, Dordrecht, p 73
347. Obenaus P, Herrmann M (1990) Pract Met 27: 503
348. Kawashima T, Okamoto H, Yamamoto H, Kitamura A (1991) J Ceram Soc Jpn 99: 1

349. Becher PF, Hwang SL, Lin HT, Ticgs TN (1994) Microstructural contributions to the fracture resistance of silicon nitride ceramics. In: Hoffmann MJ, Petzow G (eds) Tailoring of Mechanical Properties of Si_3N_4 Ceramics, NATO ASI Ser E Vol. 276, Kluwer Academic Publishers, Dordrecht, p 87

350. Schneider JA, Mukherjee AK (1999) J Am Ceram Soc 82: 761

351. Tanaka I, Nasu S, Adachi H, Miyamoto Y, Niihara K (1992) Acta met mater 40: 1995

352. Wakai F (1994) Acta met 42: 1163

353. Wiederhorn SM, Hockey BJ, Crammer DC, Yeekley R (1993) J Mater Sci 28: 445

354. Chadwick MM, Jupp RS, Wilkinson DS (1993) J Am Ceram Soc 76: 385

355. Menon MM, Fang HAT, Wu DC, Jenkins MG, Ferber MK (1994) J Am Ceram Soc 77: 1228

356. Gogotsi YG, Grathwohl G, Thümmler F, Yaroshenko VP, Herrmann M, Taut C (1993) J Eur Ceram Soc 11 : 375

357. Bonnell DA (1986) PhD thesis, University of Michigan

358. Mandal H, Hoffmann MJ (2000) Hard and Tough α-SiALON Ceramics. In: Hampshire S, Pomeroy MJ (eds) Nitrides and Oxynitrides. Trans Tech Publications Ltd, Switzerland, p 219

359. Becher PF, Lin HAT, Hwang SL, Hoffmann MJ, Chen IW (1993) The influence of microstructure on the mechanical behavior of silicon nitride ceramics. In: Chen IW, Becher P, Mitomo M, Petzow G, Yen TS (eds) Silicon Nitride Ceramics–Scientific and Technological Advances. MRS Symposium Proc 287, MRS Pittsburgh, p 147

360. Hoffmann MJ, Schneider GA, Petzow G (1993) The potential of Si_3N_4 for thermal shock applications. In: Schneider GA, Petzow G (eds) Thermal Shock and Thermal Fatigue Behaviour of Advanced Ceramics. NATO ASI Ser E 241, Klüwer Academic Publishers, Dordrecht p 49

361. Hoffmann MJ, Petzow G (1994) Pure & Appl. Chem 66: 1807

362. Chen L, Kuy E, Groboth G (1998) Surface and Coatings Technology 320: 100–101

363. Franchini C (1981) Ceramurgia 11: 140

364. Mitomo M, Uemura Y (1981) J Mat Sci Let 16: 5527

365. Thorp JS, Sharif I (1976) J Mat Sci 11: 1494

366. Niihara K (2001) Nanomaterials and Nanocomposites with Multifunctionality. Paper presented at 25th Annual Conference on Composites, Advanced Ceramics, Materials, and Structures. Am Ceram Soc, Cocoa Beach, Fl

367. Gnesin GG, Kirilenko VM, Petrovskii VA, Gervits EI, Chernavskii YA (1982) Poroshk Metall 21: 53

368. Katz NR (1993) Applications of Silicon Nitride Based Ceramics in the U.S. In: Chen IW, Becher PF Mitomo M, Petzow G, Yen TS (eds) Silicon Nitride Ceramics– Scientific and Technological Advances. Mat Res Soc Symp Proc 287, MRS, Pittsburgh, p 197

369. Subirats M, Iskander MF, White MJ, Kiggans JO (1997) J of Microwave Power and Electromagnetic Energy 32: 1997

370. Willert-Porada M, Dhupia G, Müller G, Nagel A (1999) Material and Technology Development for Microwave Sintering of High Performance. In: Müller G (ed) Ceramics-Ceramics-Processing, Reliability, Tribology and Wear, EUROMAT-Vol 12. Wiley-VCH, Weinheim, p 88

371. Watari K Seki Y, Ishizaki K (1989) J Ceram Soc Jpn Int Ed 97: 170

372. Kitayama M, Hirao K, Toriyama M, Kanzaki S (1999) J Am Ceram Soc 82: 3105

373. Watari K, Hirao K, Toriyama M, Ishizaki K (1999) J Am Ceram Soc 82: 777

374. Kitayama M, Hirao K, Watari K, Toriyama M, Kanzaki S (2001) J Am Ceram Soc 84: 353

375. Li B, Pottier L, Roger JP, Fournier D, Watari K, Hirao K (1999) J Eur Ceram Soc 19: 1631

376. Kitayama M, Hirao K, Tsuge A, Watari K, Toriyama M, Kanzaki S (2000) J Am Ceram Soc 83: 1985

377. Hirao K, Imamura H, Watari K, Brito ME, Toriyama M, Kanzaki S (1999) Seeded Silicon Nitride: Microstructure and Performance. In: Niihara K, Sekino T, Yasuda E,

Sasa T (eds) The Science of Engineering Ceramics II, Key Eng Mater 161–163. Trans Tech Publications, Switzerland, p 469
378. Mitomo M, Izumi F, Greil P, Petzow G (1984) Bull Am Ceram Soc 65: 84
379. Herrmann M, Janecke-Rößler K (1999) Lab Rep, IKTS Dresden
380. Shermann D, Brandon D (2000) Mechanical Properties and their Relation to Microstructure. In: Riedel R (ed) Handbook of Ceramic Hard Materials. Wiley-VCH, Weinheim, p 66
381. Munz D, Fett T (1999) Ceramics Mechanical Properties, Failure Behaviour, Materials Selection. Springer, Berlin, Heidelberg, New York
382. Quinn GD (1990) J Mater Sci 25: 4361
383. Wachtmann JB (1996) Mechanical Properties of Ceramics, John Wiley & Sons Inc, New York
384. Kondo N, Yozuuki Y, Ohji T (1999) J Am Ceram Soc 82: 1067
385. Tajima Y, Urashima K (1994) Improvement of Strength and Toughness of Silicon Nitride Ceramics. In: Hoffmann MJ, Petzow G (eds) Tailoring of Mechanical Properties of Si_3N_4 Ceramics, NATO ASI Series, Series E Vol 276. Kluwer Academic Publishers, Dordrecht, p 101
386. Tseng WJ, Kita H, (2000) Ceram Intern 26: 197
387. Matsumaru K, Ishizaki K (2001) Fabrication of Porous Materials with high Fracture Strength. In: Singh M, Jessen T (eds) 25th Annual Conference on Composites, Advanced Ceramics, Materials, and Structures: B, (Ceram Eng Sci Proc 22). Am Ceram Soc, Westerville, OH, p 197
388. Sajgalik P, Dusza J, Hoffmann MJ (1995) J Am Ceram Soc 78: 2619
389. Evans AG (1990) J Am Ceram Soc 73: 187
390. Rödel J (1992) J Eur Ceram Soc 10: 143
391. Kleebe HJ, Pezzotti G, Ziegler G, (1999) J Am Ceram Soc 82: 1857
392. Rice RW (2000) Mechanical Properties of Ceramics and Composites, Marcel Dekker, New York, p 245
393. Faber KT, Evans AG(1983) Acta Metall 31: 565 and 577
394. Emoto H, Hirotsuri H (1999) Microstructure Control of Silicon Nitride Ceramics Fabricated from α Powder Containing Fine β-Nuclei. In: Niihara K, Sekino T, Yasuda E, Sasa T (eds) The Science of Engineering Ceramics II Key Eng Mat 161–163. Trans Tech Publications, Switzerland, p 209
395. Walker T (1996) PhD Theses, Technical University of Dresden
396. Li CW, Gasdaska CJ, Goldacker J, Lui SC (1993) Damade Resistance of In Situ Reinforced Silicon Nitride. In: Chen IW, Becher PF, Mitomo M, Petzow G, Yen TS (eds) Silicon Nitride Ceramics – Scientific and Technological Advances. Mat Res Soc Symp Proc 287. Materials Research Soc, Pittsburgh, p 473
397. Urashima K, Ikea Y, IRAs S (1998) Features of Superior Strength Si_3N_4 Ceramic and its Application. In: Niihara K, Hirano S, Kanzaki S, Komeya K, Miring K (eds) 6th Int Symp Ceramic Materials and Components for Engines. Jap Fine Ceramic Ass, Tokyo, p 167
398. Silver RF, Vienna JM (1995) J Mat Sci 30: 5531
399. Peterson I, Tien T (1995) J Am Ceram Soc 78: 2345
400. Tanaka I, Pezzotti G, Okamoto T, Miyamoto Y, Koizumi M (1992) J Am Ceram Soc 74 : 752
401. Becher PF, Sun EY, Hsueh CH, Painter GS, More KL (2000) Role of Intergranular Films in Toughened Ceramics. In: Sajgalic P, Luences Z (eds) Engineering Ceramics Multifunctional Properties – New Perspectives, Key Eng Mat 175–176. Trans Tech Publications, Switzerland, p 97
402. Speicher R, Schneider GA, Dreßler W, Lindemann G, Böder H, Knoblauch V (1999) Reliability of Ceramic Valve Plates for Common Rail Injection Pumps. In: Müller G (ed) Ceramic Processing, Reliability, Tribology and Wear, EUROMAT99-Vol 12, Wiley-VCH, Weinheim, p 333
403. Ogasawara T, Mabuchi Y (1993) J Ceram Soc Jpn Int Edition 101: 1122
404. Jacobs DS, Chen IW (1994) J Am Ceram Soc 77: 1153

405. Bhatnagar A, Hoffmann MJ, Dauskardt RH (2000) J Am Ceram Soc 83: 585
406. Andrievski RA (1994) High Temp–High Press 26: 451
407. Miao H, Qi L, Cui G (1995) Key Engin Mater 114: 135
408. Rice RW, Wu CC, Borchelt F (1994) J Am Ceram Soc 77: 2539
409. Petzow G, Hoffmann MJ (1993) Mat Sci Forum 91: 113–115
410. Klemm H, Tangermann K, Schubert C, Hermel W (1996) J Am Ceram Soc 79: 429
411. Tanaka I, Pezzotti G, Matsushita K, Miyamoto Y, Okamoto T (1992) J Am Ceram Soc 74: 752
412. Luecke WE, Wiederhorn SM (1999) J Am Ceram Soc 82: 2769
413. Lofaj F, Wiederhorn SM, Long GG, Jemian PR (2001) Tensile Creep in the next Generation Silicon Nitride. In: Singh M, Jessen T (eds) 25th Annual Conference on Composites, Advanced Ceramics, Materials, and Structures: A (Ceram Eng Sci Proc 22). Am Ceram Soc, Westerville, OH, p 167
414. Ohji T (2001) Long Term Tensile at Creep Behaviours of Highly Heat-Resistant Silicon Nitride Ceramics Gas Turbines. In: Singh M, Jessen T (eds) 25th Annual Conference on Composites, Advanced Ceramics, Materials, and Structures: A (Ceram Eng Sci Proc 22). Am Ceram Soc, Westerville, OH, p 159
415. Yoshida M, Tanaka K, Tsuruzono S, Tatsuki T (1999) Development of Ceramic Components for Ceramic Gas Turbine Engine (CGT302). In: Vincentini P (ed) Ceramics: Getting Into the 2000's, Part D, Elsevier, London, p 253
416. Luecke WE, Wiederhorn SM, Hockey BJ, Krause RF, Long GG (1995) J Am Ceram Soc 78: 2085
417. Raj R, Chyung CK (1981) Acta Metall 29, 159
418. Thouless MD, Evans AG (1984) J Am Ceram Soc 67: 721
419. Ohji T, Yamauchi Y (1993) J Am Ceram Soc 76: 3105
420. Ohji T (1994) Tensile Creep Rupture And Subcritical Crack Growth of Silicon Nitride. In: Hoffmann MJ, Petzow G (eds) Tailoring of Mechanical Properties of Si_3N_4 Ceramics. Kluwer Academic Publishers, Netherlands, p 339
421. Marion JE, Evans AG, Drory MD, Clarke DR (1983) Acta Metall 31: 1445
422. Dryden JR, Kucerovsky D, Wilkinson DS, Watt DF (1989) Acta Metall 37: 2007
423. Klemm H, Herrmann M, Schubert C (1998) High Temperature Oxidation of Silicon Nitride Based Ceramics Materials. In: Niihara K, Hirano S, Kanzaki S, Komeya K, Morinaga K (eds) 6th Int Symp Ceramic Materials and Components for Engines. Jap Fine Ceramic Ass, Tokyo, p 576
424. Pezzotti G, Ota K, Kleebe HJ (1997) J Am Ceram Soc 80: 2341
425. Klemm H, Pezzotti G (1994) J Am Ceram Soc 77: 553
426. Xie RJ, Mitomo M, Zhan GD (2000) Acta mater 48: 2049
427. Shigegaki Y, Inamura T, Suzuki A, Sasa T (1993) High Temperature Fatigue Properties of Silicon Nitride in Nitrogen Atmosphere. In: Chen IW, Becher PF Mitomo M, Petzow G, Yen TS (eds) Silicon Nitride Ceramics - Scientific and Technological Advances. Symp Proc 287, Mat Res Soc, Pittsburgh, p 461
428. Lin CKJ, Jenkins MG, Ferber MK (1993) Evaluation of Tensile Static, Dynamic and Cyclic Fatigue Behaviour for a HIPed Silicon Nitride at Elevated Temperatures. In: Chen IW, Becher PF, Mitomo M, Petzow G, Yen TS (eds) Silicon Nitride Ceramics – Scientific and Technological Advances. Symp Proc 287, Mat Res Soc, Pittsburgh, p 455
429. Herrmann M, Klemm H, Schubert C, Hermel W (1997) Long Term Behaviour of SiC/Si_3N_4-Nanocomposites at 1400–1500 °C. In: Baxter J, Fordham LCR, Gabis V, Hellot Y, Lefebvre M, Le Doussal H, Sech AL (eds) Euro Ceramics V Key Eng Mat 132–136. Trans Tech Publications, Switzerland, p 1997
430. Jack KH (1993) SIALON CERAMICS: Retrospect and Prospect. In: Chen IW, Becher PF, Mitomo M, Petzow G, Yen TS (eds) Silicon Nitride Ceramics – Scientific and Technological Advances. Mat Res Soc Symp Proc 287. Materials Research Soc, Pittsburgh, p 15

431. Opila EJ, Jacobson NS (2000) Corrosion of ceramics. In: Schütze M (ed) Corrosion and Environmental Degradation. Wiley-VCH, Weinheim, p 329
432. Gogotsi YG, Lavrenko VA (1992) Corrosion of High Performance Ceramics. Springer, Berlin, Heidelberg, New York
433. Jacobson NS (1993) J Am Ceram Soc 76: 3
434. Nickel KG (ed) (1994) Corrosion of Advanced Ceramics – Measurement and Modelling. Kluwer Academic Publishers, Dordrecht
435. Fordham RJ, Baxter DJ, Graziani T (eds) (1996) Corrosion of Advanced Ceramics, Key Eng Mater 113. Trans Tech Publications, Switzerland
436. Nickel KG, Gogotsi YG (2000) Corrosion of Hard Materials. In: Riedel R (ed) Handbook of Ceramic Hard Materials. Wiley-VCH, Weinheim, p 140
437. Telle R, Quirnbach P (eds) (1994) Korrosion und Verschleiß von Keramischen Werkstoffen: Kombinierte Belastungsfälle als Anwendungsgrenze. DGM Informationsgesellschaft, Oberursel
438. Richter HJ, Kaiser A, Taut C, Herrmann M (1994) Anwendungsmöglichkeiten und Grenzen von Thermodynamischen Berechnungen bei der Interpretation und Voraussage von Korrosionsprozessen. In: Telle R, Quirnbach P (eds) Korrosion und Verschleiß von Keramischen Werkstoffen: Kombinierte Belastungsfälle als Anwendungsgrenze. DGM Informationsgesellschaft, Oberursel, p 322
439. Wagner C (1958) J Appl Physics 29: 1295
440. Herrmann M, Schober R (1989) Silikattechnik 40: 260
441. Opila EJ, Smialek JL, Robinson RC, Fox DS, Jacobson NS (1999) J Am Ceram Soc 82: 1826
442. Filsinger D, Schulz A, Wiitig S, Klemm H, Taut C, Wötting G (2000) Model Cobustor to Assess the Oxidation Behavior of Ceramic Materials under Engine Conditions. In: Nelson HD (ed) Internat Gas Turbine & Aeroengine Congress & Exhibition 2000: Transactions of the ASME, ASME, New York, p 2000-GT-349
443. Newson DD, Pollinger JP, Twait DJ (2000) Internat Gas Turbine & Aeroengine Congress & Exhibition 2000: Transactions of the ASME, New York, p 2000-GT-533
444. Yokoyama K, Wada S (2000) J Ceram Soc Jpn 108: 627
445. Klemm H, Herrmann M, Schubert C (2000) J Eng. Gas Turbines and Power 122: 13
446. Klemm H, Herrmann M, Schubert C (1996) The Influence of the Grain Boundary Phase Composition on the High Temperature Properties of Silicon Nitride Materials. In: Parilak L, Danninger H, Dusza J, Weiss B (eds) Proc Int Conf Deformation and Fracture in Structural PM Materials, IMR SAS Kosice 2: 75
447. Andrews P, Riley FL (1991) J Eur Ceram Soc 7: 125
448. O'Meara C, Sjödberg J (1997) J Am Ceram Soc 80: 1491
449. Ricoult-Barkhausen M, Gogotsi YG (1996) Identification of Oxidation Mechanisms in Silicon Nitride Ceramics by TEM. In: Fordham PJ, Baxter DJ, Graziani T (eds) Corrosion of Advanced Ceramics. Key Eng Mat 113. Trans Tech Publications, Switzerland, p 81
450. Luthra KL (1994) Theoretical Aspects of the Oxidation of Silica-Forming Ceramics. In: Nickel KG (ed) Corrosion of Advanced Ceramics – Measurement and Modelling. Kluwer Academic Publishers, Dordrecht, p 23
451. Luthra KL (1991) J Am Ceram Soc 74: 1095
452. Ogbuji LUJT (1994) The Oxidation Process in Silicon Nitride. In: Nickel KG (ed) Corrosion of Advanced Ceramics – Measurement and Modelling. Kluwer Academic Publishers, Dordrecht, p 117
453. Ogbuji LUJT, Bryan SR (1995) J Am Ceram Soc 78: 1272
454. Ogbuji LUJT, Opila EJ (1995) J Electrochem Soc 142: 925
455. Taut C (1994) PhD thesis, University of Dresden
456. Tressler RE (1990) Environmental Effects on Long Therm Reliability of SiC and Si_3N_4. In: Tressler RE, McNallan M (eds) Ceramic Transactions 10, Am Ceram Soc, Westerville, OH, p 99

457. Van der Biest O, Weber C (1994) Influence of Oxidation on Long Term Reliability of Silicon Nitride. In: Nickel KG (ed) Corrosion of Advanced Ceramics – Measurement and Modelling. Kluwer Academic Publishers, Dordrecht, p 453

458. Nishimura N, Masuo E, Takita K (1991) Effect of Microstructural Oxidation on the Strength of Silicon Nitride after High Temperature Exposure. In: Carlsson T, Johansson T, Kahlman T (eds) Proc 4th Int Symp Ceram Mater & Components for Engines. Elsevier, London, p 1139

459. Mukundhan P, Wu P, Du HH (1999) J Am Ceram Soc 82: 226

460. Herrmann M, Klemm H, Göbel B, Schubert C, Hermel W (1999) SiC/Si3N4 Nanocomposites with Excellent High Temperature Long-Term Behaviour. In: Niihara, Sekino, Yasuda E, Sasa T (eds) The Science of Engineering Ceramics II, Key Eng Mat 161–163. Trans Tech Publications, Switzerland, p 377

461. Watanabe M, Shimamori T, Noda Y (1992) The High Temperature Properties of Sc_2O_3 doped Si_3N_4. In: Ishizaki K (ed) Grain Boundary Controlled Properties of Fine Ceramics, Elsevier, London, p 199

462. Nordberg LO, Käll PO, Nygren M (1996) A Mathematical Analysis of the Non-Parabolic Oxidation Behaviour of β-SiALON Matrices and Composites. In: Fordham PJ, Baxter DJ, Graziani T (eds) Corrosion of Advanced Ceramics, Key Eng Mat 113. Trans Tech Publications, Switzerland, p 39

463. Patel JK, Thompson DP (1988) Br Ceram Trans J. 87: 70

464. Baxter DJ, Graziani T, Wang HM, McCauley RA (1998) J Eur Ceram Soc 18: 2323

465. Ernstberger U (1985) PhD Thesis, University of Karlsruhe

466. Ernstberger U, Grathwohl G, Thümmler F (1987) Int J High Techn Ceram 3: 43

467. Babini GN, Belosi A, Vincenzini P (1983) J Mater Sci 18: 231

468. Loehman RE (1989) Am Ceram Soc Bull 68: 891

469. Loehman RE (1999) Key Engin Mat 657: 161–163

470. Tönshoff HK, Denkena B (1991) J Society of Tribologists and Lubrication Engineers 47: 772

471. Miao H, Qi, L, Cui G (1995) Silicon Nitride Ceramic Cutting-Tools and their Applications. In: Low IM, Li XS (eds) Advanced Ceramic Tools for Machining Application – II Key Engineering Materials 114, Trans Tech Publications Ltd, Switzerland, p 135

472. Voitovich RF (1971) Tugoplavkie Soedineniya, termoddinamicheskie Kharakteriustiki, Naukova Duma, Kiev

473. Samsonov GV, Vinitskii IM (1980) Handbook of Refractory Compounds, Plenum, New York, p 400

474. Feld G (1969) Sprechsaal 102: 1098

475. Watanabe M, Usami T, Takasu S, Matsuo S, Toji E (1983) Jap J Appl Phys 1 22: 185

476. Li JG, Hausner H (1992) J Eur Ceram Soc 9: 101

477. Li JG, Hausner H (1990) Benetzung von anorganischen Materialien durch Siliciumschmelzen, Abschlußbericht, TU Berlin, p 37

478. Swartz JC (1976) J Am Ceram Soc 59: 272

479. Singh RN, Tuohig WD (1975) J Am Ceram Soc 58: 70

480. Smith PL, White J (1983) Trans J Br Ceram Soc 82: 23

481. Belyi VI (1988) Mater Sci Mnogr 34: 126

482. David J (1972) Rev Chim Miner 9: 717

483. Leimer G, Gugel E (1975) Z Metallkd 66: 570

484. Panacjuk AD, Fomenko VS, Glebova GG (1986) Stoikost nemetallischeskich materialov b rasplavach. Naukova Dumka, Kiev, p 175

485. Jelacic C, Dervisbegovic H (1974) Bull Soc Fr Ceram 105: 17

486. Felten RP (1974) Sprechsaal 107: 92, 101

487. Naka M (1991) Ultramicroscopy 39: 128

488. Naka M, Kubo M, Mori H, Okamoto I (1990) Koon Gakkaishi 16: 225

489. Gilde W (1953) Metall Giessereitech 3: 324

490. Schwabe U, Wolff L, Ziegler G (1988) Klei/Glas/Keram 9: 231

491. Feld H, Gugel E, Nitzche HG (1969) Werkst Korros 20: 571
492. Ljungberg L, Warren R (1989) Ceram Eng Sci Proc 10: 1655
493. Brunken R (1970) Werkstatt Betr 103(1): 65
494. Schuster JC, Weitzer F, Bauer J, Nowotny H (1988) Mater Sci Eng A 105/106: 201
495. Weitzer F, Schuster JC, Bauer J, Jounel B (1991) J Mater Sci 26: 2076
496. Naidich YV, Zhuravlev VS, Frumina NI, Kostyuk BD, Krasovskaya NA Ostrovskii VG (1988) Poroshk Metall 11: 58
497. Kunz KP, Sarin VK, Davis RF Bryan SR (1988) Mater Sci Eng A 105/106: 47
498. Dummler W, Weber S, Tete C, Scherrer S (1999) J Mater Sci Lett 18: 193
499. Akselsen OM (1992) J Mat Sci 27: 1989
500. Raic KT (1999) Ceram Int 26/1: 19
501. Shimoo T, Okamura K, Shibata D (2000) J Mater Sci 35/21: 5485
502. Schuster JC (1988) J Mat Sci 23: 2792
503. Glemser O, Beltz K, Naumann P (1957) Z Anorg Allg Chem 291: 51
504. Shimoo T, Okamura K, Yamasaki T (1999) J Mater Sci 34: 5525
505. Weitzer F, Schuster JC (1987) J Solid State Chem 70: 178
506. Komeya K, Meguro T, Atago S, Lin CH, Abe Y, Komatsu M (1999) Corrosion Resistance of Silicon Nitride Ceramics. In: Niihara K, Sekino T, Yasuda E, Sasa T (eds) The Science of Engineering Ceramics II. Key Eng Mat 161–163, Trans Tech Publications, Switzerland, p 235
507. Hollstein T, Graas T, Bundschuh K, Schütze M (1998) Keram Zeitschrift 50: 416
508. Sato T, Tokunaga Y, Endo T, Shimada M, Komeya K (1988) J Mat Sci 23: 3440
509. Iio S, Okada A, Akira A, Tetsuo A, Masahiro Y (1992) J Ceram Soc Jpn Int Edition 100: 954
510. Herrmann M, Schubert C, Michael G (1999) Korrosionsstabile keramische Werkstoffe für Anwendungen in Wälzlagern und im Anlagenbau. In: cfi DKG Berichte 14: 130
511. Herrmann M, Michael G (1999) Br Ceram Proc 60/1: 455
512. Okada A, Yoshimura M (1996) Mechanical Degradation of Silicon Nitride Ceramics in Corrosive Solutions of Boiling Sulphuric Acid. In: Fordham RJ, Baxter DJ, Graziani T (eds) Corrosion of Advanced Ceramics, Key Eng Mater 113. Trans Tech Publications, Switzerland, p 227
513. Gogozi YG, Lavrenko VA (1992) Springer Verlag, Berlin, p 76
514. Shimada M, Sato T (1989) Ceram Trans 10: 355
515. Fang Q, Sidky PS, Hocking MG (1997) Corr Sci 39: 511
516. Kanbara K, Uchida N, Uematsu K, Kurita T, Yoshimoto K, Suzuki Y (1993) Corrosion of Silicon Nitride Ceramics by Nitric Acid. In: Chen IW, Becher PF, Mitomo M, Petzow G, Yen TS (eds) Silicon Nitride Ceramics, Mat Res Soc Symp Proc 287. Mat Res Soc, Pittsburgh, p 533
517. Wötting G, Herrmann M, Michael G, Siegel S, Frassek L (1999) WO99/20579
518. Sharkawy SW, El-Aslabi AM (1998) Corros Sci 40: 1119
519. Sato T, Tokunaga Y, Endo T, Shimada M, Komeya K, Nishida K, Komatsu M, Kameda T (1988) J Mat Sci 23: 3440
520. Sato T, Murakami T, Shimada E, Komeya K, Komeda T, Komatsu M (1991) J Mat Sci 26: 1749
521. Yoshimura M, Yamamoto S (1995) Seramikkusu 30: 995
522. Oda K, Yoshio T, Miyamoto Y, Koizumi M (1993) J Am Ceram Soc 76: 1365
523. Hollstein T (1999) Adv Sci Technol 13: 433
524. Sato T, Sato S, Tamura K, Okuwaki A (1992) J Br Ceram Trans 91: 117
525. Herrmann M, Michael G, Schubert C, Adler J, Krell A (1998) Final report MATECH project 03N2002 F
526. Ye J, Furuya K, Misono Y, Matsuo K, Munakata F, Ishikawa I, Akimune Y (2000) J Luminescence 574: 87–89
527. Shen Z, Nygren M, Halenius U (1997) J Mat Sci Lett 16: 263
528. Pechenik A, Piermarini GJ (1992) J Am Ceram Soc 75: 3283
529. Rouxel T, Piriou B (1996) J Appl Phys 79: 9074

530. Keßler S (1993) PhD Thesis, University of Stuttgart
531. Clarke DR (1989) Mat Sci. Forum 47: 110
532. Yamakawa A, Miyake M, Ishizaki K (1994) J Ceram Soc Jpn 102: 339
533. Lewis HM (1993) Sialons and Silicon Nitrides; Microstructural Design and Performance. In: Chen IW, Becher PF, Mitomo M, Petzow G, Yen TS (eds) Silicon Nitride Ceramics – Scientific and Technological Advances, Mat Res Soc Symp Proc 287, MRS, Pittsburgh, p 159
534. Kessler H, Herrmann M, Pompe W (1995) Acta metall mater 43: 2789
535. Keßler H (1995) PhD thesis, Technical University of Dresden
536. Carborundum Co. (1952) US Patent No 2618565
537. Moulson AJ (1979) J Mat Sci 14: 1017.
538. Heinrich JG, Krüner H (1994) Silicon Nitride Materials for Engine Applications. In: Hoffmann MJ, Petzow G (eds) Tailoring of Mechanical Properties of Si_3N_4 Ceramics. NATO ASI Series E, Vol 276, Kluwer Academic Publishers, Dordrecht, p 19
539. Ziegler G (1985) Z Werkstofftechnik 16: 12, 45, 81
540. Mikijeli B, Mangels J (2001) SRBSN Material Development For Automotive Applications. In: Heinrich JG, Aldinger F (eds) Ceramic Materials Components for Engines, Wiley-VCH, Weinheim, p 393
541. Riley FF (1983) Silicon Nitridation. In: Riley FF (ed) Progress in Nitrogen Ceramics. Martinus Niihoff Publishers, Dordrecht, p 121
542. Jennings HM, Richmann MH (1976) J Mat Sci 11: 285
543. Ziegler G, Heinrich JG, Wötting G (1987) J Mat Sci 22: 3041
544. Rahaman MN, Moulson AJ (1984) J Mat Sci 19: 189
545. Jennings M (1983) J Mat Sci 18: 951
546. Atkinson A, Moulson AJ, Roberts EW (1976) J Am Ceram Soc 59: 285
547. Rossetty GA, Denkewicz RP (1989) J Mat Sci 24: 3081
548. Li WB, Lei BQ, Lindback T (1997) J Eur Ceram Soc 17: 1119
549. Campos-Loriz D, Riley FL (1978) J Mat Sci 13: 1125
550. Sheldon BW, Rankin J, Haggerty JS (1995) J Am Ceram Soc 78: 1624
551. Mangels JA, Tennenhouse GJ (1980) Am Ceram Soc Bull 59: 1216
552. Herrmann M, Heß S, Pabst S, Richter HJ, Hermel W (1990) Keram Zeitschr 42: 250
553. Wötting G (1999) Final report BMBF- project 03N2002B
554. Liu H, Hsu SM (1996) J Am Ceram Soc 79: 2452
555. Schober R, Richter HJ (1999) Composite ceramics based on silicon nitride or aluminium nitride. In: Krams J (ed) Proceedings Biennial Worldwide Congress UNITECR '99, Verlag Stahleisen, Düsseldorf, p 198
556. Isomura K, (1990) J Am Ceram Soc 73: 624
557. Petrovsky VY, Rak ZS (2001) J Eur Ceram Soc 21: 219
558. Petrovsky VY, Rak ZS (2001) J Eur Ceram Soc 21: 237
559. Gogotsi YG (1994) J Mat Sci 29: 2541
560. Sigulinski F, Boskovic S (1999) Ceram Intern 25: 41
561. Liu CC (2000) J Ceram Soc Jpn 108: 46
562. Yamada K, Kamiya N (1999) Mat Sci Eng 261: 270
563. Herrmann M, Bales A, Bach E, Leuteritz U (unpublished)
564. Kawamura H (1999) New Perspectives in Engine Applications of Engineering Ceramics. In: Niihara K, Sekino T, Yasuda E, Sasa T (eds) The Science of Engineering Ceramics II, Key Eng Mat 161–163. Trans Tech Publications, Switzerland, p 9
565. Veprek S (2000) Nanostructured Superhard Materials. In: Riedel R (ed) Handbook of Ceramic Hard Materials. Wiley VCH, Weinheim, p 104
566. Pezzotti G, Tanaka I, Ikuhara Y, Sakai M, Nishida T (1994) Scripta Met Et Mat 31: 403
567. Seifert HJ (1993) PhD Thesis, University of Stuttgart
568. Herrmann M, Balzer B, Schubert C, Hermel W(1993) J Eur Ceram Soc 12: 287
569. Huang JL, Chiu HL, Lee MT (1994) J Am Ceram Soc 77: 705
570. Herrmann M, Schubert C, Hermel W, Meißner E, Ziegler G (1996) cfi/Ber DKG 73: 434
571. Fricke M, Nonninger R, Schmidt H (2000) Adv Eng Mat 2(10): 647

572. Lee BT, Yoon YJ, Lee KH (2001) Mater Lett 47: 71
573. Wang CM (1995) J Mater Sci 30: 3222
574. Iwamoto Y, Kikuta K, Hirano S (2000) J Ceram Soc Jpn 108(4): 350 and 1072
575. Nagaoka T, Yasuoka M, Hirao K, Kanzaki S (1992) J Ceram Soc Jpn 100: 612
576. Bellosi A, Guicciardi S, Tampieri A (1992) J Eur Ceram Soc 9: 83
577. Liu CC, Huang JL (2000) Brit Ceram Trans 99: 149
578. Akimune Y, Munakata F, Hirosaki N, Okamoto Y (1998) J Ceram Soc Jpn 106: 75
579. Opsommer A, Gomez E, Castro F (1998) J Mater Sci 33: 2583
580. Woydt M, Skopp A, Habig KH (1991) Wear 148: 377
581. Skopp A (1993) PhD Thesis, University of Berlin
582. Xu J, Kato K (1997) Wear 202: 165
583. Popp M, Sternagel R, Pfeifer W, Blug B, Meier S, Wötting G, Frasseck L (2000) Hybrid- and Ceramic Rolling Bearings with Modified Surface and Low friction Rolling Contact. In: Müller G (ed) Ceramics – Processing, Reliability, Triboloy and Wear, Euromat 99 Vol 12. Wiley VCH, Weinheim, p 449
584. Lences Z, Sajgalik P, Toriyama M (2000) J Eur Ceram Soc 20: 347
585. Yeh CH, Hon MH (1996) Ceram Intern 23: 361
586. Lange FF (1973) J Am Ceram Soc 56: 445
587. Herrmann M, Schubert C, Klemm H (1996) Nanocompositkeramik: Übersicht über Werkstoffkonzepte, Herstellungstechnologien und spezifische Eigenschaften. In: Kriegesmann J (ed) Technische Keramische Werkstoffe, Deutscher Wirtschaftsdienst, Köln, Chap 4.4.3.0.
588. Greil P, Petzow G, Tanaka H (1987) Ceram Int 13: 19
589. Hockey BJ, Widerhorn SM, Liu W, Balfdoni G, Buljan ST (1991) J Mater Sci 26: 3931
590. Kodama H, Sakamoto H, Miyoshi T (1989) J Am Ceram Soc 72: 551
591. Laughner JW, Bhatt RT (1989) J Am Ceram Soc 72: 2017
592. Becher PF, Hsueh CH, Angelini P, Tiegs TN (1988) J Am Ceram Soc 71: 1050
593. Campbell GH, Rühle M, Dalgleish BJ, Evans AG (1990) J Am Ceram Soc 73: 521
594. Birchall JD, Stanley DR, Mockford MJ, Pigott GH, Pinto PJ (1988) J Mat Sci Lett 7: 350
595. Pezzoti G (1993) J Am Ceram Soc 76: 1313
596. Rendtel A, Hübner H, Schubert C (1995) Silicates Industriels 60: 305
597. Niihara K, Izaki K, Kawakami T (1990) J Mat Sci Lett 10: 112
598. Sternitzke M (1997) J Eur Ceram Soc 17: 1061
599. Izaki K, Hakkei K, Ando K, Kawakami KT, Niihara K (1988) Fabrication and Mechanical Properties of Si_3N_4/SiC Composites from Fine, Amorphous Si-C-N- Powder Precursors. In: Mackenzie D (ed) Ultrastructural Processing of Advanced Ceramics, John Wiley & Sons Inc, New York, p 891
600. Hirano T, Niihara K (1995) Mater Lett 22: 249; (1996) 26: 285
601. Lavedrine A, Bahloul D, Coursat P (1991) J Eur Ceram Soc 8: 221
602. Palchevskis E, Grabis J, Millers T (1993) Fine Si_3N_4-SiC powders; Obtaining and Characteristics. In: Aldinger F (ed) PTM '93, DGM Informationsgesellschaft, Oberursel, p 657
603. Cauchetier M, Croix O, Luce M (1991) J Eur Ceram Soc 8: 215
604. Westerheide R, Wöting G, Schmitz HW (1998) SiC-Si_3N_4 Nanocomposite- Potential und wirtschaftliche Realisierung. In: Kriegesmann I (ed) Keramische Werkstoffe 84erg. -lfg. Deutscher Wirtschaftsdienst, Köln, Chap 4.4.3.2
605. Wötting G, Caspers B, Gugel E, Westerheide R (2000) Transactions of the ASME 122: 8
606. Ishizaki K, Yanai T (1995) Silicates Industriels 60: 215
607. Sajgalik P, Hnatko M, Lences Z, Warbichler P, Hofer F (2001) Z Metallkd 92: 937
608. Riedel R, Klebe HJ, Schönfelder H, Aldinger F (1995) Nature 374: 526
609. Riedel R, Seher M, Mayer J, Szabo DV (1995) J Eur Ceram Soc 15: 703 and 717
610. Haluschka C, Kleebe HJ, Franke R, Riedel R (2000) J Eur Ceram Soc 20: 1355
611. Haluschka C, Engel C, Riedel R (2000) J Eur Ceram Soc 20: 1365
612. Sajgalik P, Hnatko M, Lofai F, Hvizdos P, Dusza J, Warbichler P, Hofer F, Riedel R, Lecomte E, Hoffmann MJ (2000) J Eur Ceram Soc 20: 453

613. Ukyo Y, Kandori T, Wada S (1993) J Ceram Soc Jpn Int Ed 101: 1398
614. Niihara K, Suganuma K, Nikahira K, Izaki K (1990) J Mater Sci Lett 9: 598
615. Pan X, Mayer J, Rühle M, Niihara K (1994) J Am Ceram Soc 77: 3039
616. Cheong DS, Hwang KT, Kim CS (1999) J Am Ceram Soc 82: 981
617. Rouxel T, Wakai F (1992) J Am Ceram Soc 75: 2363
618. Rendtel A, Hübner H, Herrmann M, Schubert C (1998) J Am Ceram Soc 81: 1109
619. Ramoul-Badabache K, Lancin M (1992) J Eur Ceram Soc 10: 369
620. Klemm H, Tangermann K, Reich T, Herrmann M, Schubert C, Hermel W (1995) High-Temperature Properties of Silicon Nitride Molybdenum Silicide Composites. In: Bellosi A (ed) Fourth Europ Ceramics, Vol 4, Grupp Editoriale faenca Editrice, Faenza, p 233
621. Pezotti G, Sakai M (1994) J Am Ceram Soc 77: 3039
622. Rendtel A, Hübner H, Herrmann M (1995) Creep Behaviour of Si_3N_4/SiC-Nanocomposite Materials. In: Bellosi A (ed) Fourth Europ Ceramics Vol 4, Grupp Editoriale faenca Editrice S p.A, Faenza, p 225
623. Klebe HJ, Pezzotti G, Rühle M (1998) J Ceram Soc Jpn 106: 17
624. Rendtel P, Rendtel A Hübner H, Herrmann M (1999) J Eur Ceram Soc 19: 217
625. Gommes JR, Osendi MI, Miranzo P, Oliveira FJ, Silva RF (1999) Wear 222: 233–235
626. Sawagushi A, Toda K, Niihara K (1991) J Am Ceram Soc 74: 1142
627. Kanzaki S, Brito ME, Valecillos MC, Hirao K, Toriyama M (1997) J Eur Ceram Soc 17: 1841
628. Akimune Y, Munakata F, Matsuo K, Okamoto Y, Hirosaki N, Satoh C (1999) J Ceram Soc Jpn 107: 1180
629. Imamura H, Hirao K, Brito ME, Toriyama M, Kanzaki S (2000) J Am Ceram Soc 83: 495
630. Hirao K, Nagaoka T, Brito ME, Kanzaki S (1994) J Am Ceram Soc 77: 1857
631. Teshima H, Hirao K, Toriyama M, Kanzaki S (1999) J Ceram Soc Jpn 107: 1216
632. Hirao K, Ohashi M, Brito ME, Kanzaki S (1995) J Am Ceram Soc 78: 1687
633. Lee SY, Amoako-Appiagyei K, Kim HD (1999) J Mater Res 14: 178
634. Hirosaki N, Okamoto Y, Munakata F, Akimune Y (1999) J Eur Ceram Soc 19: 2183
635. Kitayama M, Hirao K, Toriyama M, Kanzaki S (2000) J Ceram Soc Jpn 108: 646
636. Jack KH (2000) Nitrogen Ceramics for Engine Applications. In: Hampshire S, Pomoroy MJ (eds) Nitrides and Oxynitrides, Mat Sci Forum 325–326. Trans Tech Publications, Switzerland, p 255
637. Katz NR (1997) Industrial Ceram 17: 158
638. Savitz M (1999) Am Bull Ceram Soc 78 No. 1: 53
639. Brandt G (2001) Ceramic Cutting Tools. In: Heinrich JG, Aldinger F (eds) Ceramic Materials and Components for Engines. Wiley-VCH, Weinheim, p 21
640. Bayer O, Streit E (2000) Keram Zeit 52: 1092
641. Cundill RT (1997) Impact Resistance of Silicon Nitride Balls. In: Niihara K, Hirano S, Kanzaki S, Komeya K, Morinaga K (eds) Ceramic Materials and Components for Engines. Jap Fine Ceramic Assoc, Tokyo, p 556
642. Wagemann A (2000) Keram Zeit 52: 503
643. Guangchuan L, Jinsheng L, Xingua G (1999) Industrial Ceram 19: 17
644. Gugel E, Wötting G (1999) Industrial Ceram 19: 196
645. Woetting G, Frassek L, Leimer G, Schönfelder L (1993) cfi/Ber DKG 70: 287
646. Mandler WF (2001) Commercial Applications for Advanced Ceramics in Diesel Engines. In: Singh M, Jessen T (eds) 25th Annual Conference on Composites, Advanced Ceramics, Materials, and Structures: A (Ceram Eng Sci Proc 22). Am Ceram Soc, Westerville, OH, p 3
647. Mandler WF (1997) Ceramic Successes in Diesel Engines. In: Niihara K, Hirano S, Kanzaki S, Komeya K, Morinaga K (eds) Ceramic Materials and Components for Engines. Jap Fine Ceramic Assoc, Tokyo, p 137
648. Kamo R, Mavinahally NS, Kamo L, Bryzik W, Reid M (1997) Experimental Heat Release of Insulated Turbocharged Diesel Engine. In: Niihara K, Hirano S, Kanzaki S, Komeya K, Morinaga K (eds) Ceramic Materials and Components for Engines. Jap Fine Ceramic Assoc, Tokyo, p 146

649. Wötting G, Hennicke J, Feuer H, Thiemann KH, Vollmer D, Fechter E, Sticher F, Geyer A (2001) Reliability and Reproducibility of Silicon Nitride Valves: Experiences of a Field Test. In: Heinrich JG, Aldinger F (eds) Ceramic Materials and Components for Engines. Wiley VCH, Weinheim, p 181
650. Savitz M (1999) Bull Am Ceram Soc 78 No. 3: 52
651. Meiser M (2001) Current Status of Structural Ceramics. Paper presented at the 25th Annual Conference on Composites, Advanced Ceramics, Materials, and Structures. Am Ceram Soc, Cocoa Beach, Fl
652. Matsuura T, Kawai C, Yamakawa A (1997) Sumitomo Electric Technical Review 43: 77
653. Kawai C, Matsuura T, Yamakawa A (1999) J Mater Sci 43: 893
654. Kawamura H (2001) Practical use of Ceramic Components and Ceramic Engines. In: Heinrich JG, Aldinger F (eds) Ceramic Materials and Components for Engines. Wiley-VCH, Weinheim, p 27
655. Yoshida M, Tanaka K, Tsuruzono S, Tatsumi T (1999) Industrial Ceram 19: 188
656. Der Kochherd der Zukunft arbeitet mit Si_3N_4-Platten (1999) In: cfi/Ber DKG, 76, No.4: D18-20
657. Aberle AG (2001) Solar Energy Materials & Solar Cells 65: 239
658. Tsukumura K, Kamo K (2000) J Ceram Soc Jpn 108: 882

Subject Index

Author Index Volumes 101–102

Aldinger F, see Seifert HJ (2002) *101*:1–58

Frühauf S, see Roewer G (2002) *101*:59–136

Haubner R, Wilhelm M, Weissenbacher R, Lux B (2002) Boron Nitrides – Properties, Synthesis and Applications. *102*:1–46
Herrmann M, see Petzow G (2002) *102*:47–166
Herzog U, see Roewer G (2002) *101*:59–136

Jansen M, Jäschke B, Jäschke T (2002) Amorphous Multinary Ceramics in the Si-B-N-C System. *101*:137–192
Jäschke B, see Jansen M (2002) *101*:137–192
Jäschke T, see Jansen M (2002) *101*:137–192

Lux B, see Haubner R (2002) *102*:1–46

Müller E, see Roewer G (2002) *101*:59–136

Petzow G, Hermann M (2002) Silicon Nitride Ceramics. *102*:47–166

Roewer G, Herzog U, Trommer K, Müller E, Frühauf S (2002) Silicon Carbide – A Survey of Synthetic Approaches, Properties and Applications. *101*:59–136

Seifert HJ, Aldinger F (2002) Phase Equilibria in the Si-B-C-N System. *101*:1–58

Trommer K, see Roewer G (2002) *101*:59–136

Weissenbacher R, see Haubner R (2002) *102*:1–46
Wilhelm M, see Haubner R (2002) *102*:1–46

Printed in the United Kingdom
by Lightning Source UK Ltd.
124204UK00004B/157/A